AF677061

Cultural Technologies

Routledge Research in Cultural and Media Studies

For a full list of titles in this series, please visit www.routledge.com.

6 **The Nationwide Television Studies**
David Morley and Charlotte Bronsdon

7 **The New Communications Landscape**
Demystifying Media Globalization
Edited by Georgette Wang, Jan Servaes, and Anura Goonasekera

8 **Media and Migration**
Constructions of Mobility and Difference
Edited by Russell King and Nancy Wood

9 **Media Reform**
Democratizing the Media, Democratizing the State
Edited by Monroe E. Price, Beata Rozumilowicz, and Stefaan G. Verhulst

10 **Political Communication in a New Era**
Edited by Gadi Wolfsfeld and Philippe Maarek

11 **Writers' Houses and the Making of Memory**
Edited by Harald Hendrix

12 **Autism and Representation**
Edited by Mark Osteen

13 **American Icons**
The Genesis of a National Visual Language
Benedikt Feldges

14 **The Practice of Public Art**
Edited by Cameron Cartiere and Shelly Willis

15 **Film and Television After DVD**
Edited by James Bennett and Tom Brown

16 **The Places and Spaces of Fashion, 1800–2007**
Edited by John Potvin

17 **Communicating in the Third Space**
Edited by Karin Ikas and Gerhard Wagner

18 **Deconstruction After 9/11**
Martin McQuillan

19 **The Contemporary Comic Book Superhero**
Edited by Angela Ndalianis

20 **Mobile Technologies**
From Telecommunications to Media
Edited by Gerard Goggin and Larissa Hjorth

21 **Dynamics and Performativity of Imagination**
The Image between the Visible and the Invisible
Edited by Bernd Huppauf and Christoph Wulf

22 **Cities, Citizens, and Technologies**
Urban Life and Postmodernity
Paula Geyh

23 **Trauma and Media**
Theories, Histories, and Images
Allen Meek

24 **Letters, Postcards, Email**
Technologies of Presence
Esther Milne

25 **International Journalism and Democracy**
Civic Engagement Models from Around the World
Edited by Angela Romano

26 **Aesthetic Practices and Politics in Media, Music, and Art**
Performing Migration
Edited by Rocío G. Davis, Dorothea Fischer-Hornung, and Johanna C. Kardux

27 **Violence, Visual Culture, and the Black Male Body**
Cassandra Jackson

28 **Cognitive Poetics and Cultural Memory**
Russian Literary Mnemonics
Mikhail Gronas

29 **Landscapes of Holocaust Postmemory**
Brett Ashley Kaplan

30 **Emotion, Genre, and Justice in Film and Television**
E. Deidre Pribram

31 **Audiobooks, Literature, and Sound Studies**
Matthew Rubery

32 **The Adaptation Industry**
The Cultural Economy of Literary Adaptation
Simone Murray

33 **Branding Post-Communist Nations**
Marketizing National Identities in the "New" Europe
Edited by Nadia Kaneva

34 **Science Fiction Film, Television, and Adaptation**
Across the Screens
Edited by J. P. Telotte and Gerald Duchovnay

35 **Art Platforms and Cultural Production on the Internet**
Olga Goriunova

36 **Queer Representation, Visibility, and Race in American Film and Television**
Melanie E.S. Kohnen

37 **Artificial Culture**
Identity, Technology, and Bodies
Tama Leaver

38 **Global Perspectives on Tarzan**
From King of the Jungle to International Icon
Edited by Annette Wannamaker and Michelle Ann Abate

39 **Studying Mobile Media**
Cultural Technologies, Mobile Communication, and the iPhone
Edited by Larissa Hjorth, Jean Burgess, and Ingrid Richardson

40 **Sport Beyond Television**
The Internet, Digital Media and the Rise of Networked Media Sport
Brett Hutchins and David Rowe

41 **Cultural Technologies**
The Shaping of Culture in Media and Society
Edited by Göran Bolin

Cultural Technologies

The Shaping of Culture in Media and Society

Edited by Göran Bolin

Routledge
Taylor & Francis Group
NEW YORK LONDON

NEIL HELLMAN LIBRARY
COLLEGE OF ST. ROSE
ALBANY, NEW YORK

302.23
C9686b

First published 2012
by Routledge
711 Third Avenue, New York, NY 10017

Simultaneously published in the UK
by Routledge
2 Park Square, Milton Park, Abingdon, Oxon OX14 4RN

Routledge is an imprint of the Taylor & Francis Group, an informa business

© 2012 Taylor & Francis

The right of Göran Bolin to be identified as the author of the editorial material, and of the authors for their individual chapters, has been asserted in accordance with sections 77 and 78 of the Copyright, Designs and Patents Act 1988.

All rights reserved. No part of this book may be reprinted or reproduced or utilised in any form or by any electronic, mechanical, or other means, now known or hereafter invented, including photocopying and recording, or in any information storage or retrieval system, without permission in writing from the publishers.

Trademark Notice: Product or corporate names may be trademarks or registered trademarks, and are used only for identification and explanation without intent to infringe.

Library of Congress Cataloging-in-Publication Data

Cultural technologies : the shaping of culture in media and society / edited by Göran Bolin.
p. cm. — (Routledge research in cultural and media studies)
Includes bibliographical references and index.
1. Technology—Sociological aspects. 2. Technology and civilization. 3. Mass media and technology. 4. Mass media and culture. I. Bolin, Göran
HM846.C84 2012
302.230973—dc23
2011052511

ISBN13: 978-0-415-89311-4 (hbk)
ISBN13: 978-0-203-11735-4 (ebk)

Typeset in Sabon
by IBT Global.

Contents

List of Figures ix
Foreword xi

1 Introduction: Cultural Technologies in Cultures of Technology 1
GÖRAN BOLIN

PART I
Histories of Cultural Technologies

2 The Algorithmic Turn: Photosynth, Augmented Reality and the Changing Implications of the Image 19
WILLIAM URICCHIO

3 The Compact Disc and Its Culture: Notes on Melancholia 36
ERIC W. ROTHENBUHLER

4 Peer-to-Peer File-Sharing Systems: Files, Objects, Distribution 51
MATS BJÖRKIN

5 Technostruggles and the Satellite Dish: A Populist Approach to Infrastructure 64
LISA PARKS

PART II
Epistemologies of Cultural Technologies

6 Being 'Accountable': TV Audiences and Surveillance 87
TOBY MILLER

7 Time, Space and Clouds of Information: Data Centre Discourse and the Meaning of Durability 103
PETER JAKOBSSON AND FREDRIK STIERNSTEDT

8 Search Engines in Practice: Structure and Culture in Technical Development 118
ELIZABETH VAN COUVERING

9 Technology and Epistemology: Information Policy and Desire 133
SANDRA BRAMAN

PART III
Uses of Cultural Technologies

10 Web 2.0 Technologies of the Self 153
MARIA BAKARDJIEVA AND GEORGIA GADEN

11 Virtual Technologies of Nation-State: State Administration in Second Life 170
STINA BENGTSSON

12 The Scary Promise of Technology: Developing New Forms of Audience Research 189
JOKE HERMES

List of Contributors 203
Index 207

Figures

2.1	Piazza San Marco with the Basilica, by Canaletto, 1730.	26
2.2	Photosynth image of Piazza San Marco, Venice.	28
5.1	Satellite dishes and solar panels are sold at the black market in Ulaanbaatar, Mongolia.	71
5.2	Satellite dish vendors place dishes in front of shops along the street in Istanbul.	71
5.3	Satellite dishes on an apartment building in Manipur, India.	72
5.4	Satellite dish on a home in the desert town of Palo Verde, California.	73
5.5	Al Jessup adjusting a satellite dish at his home in Bethany, West Virginia.	79
5.6	Satellite television dishes cover the home of Lee Si-kap, a farmer.	80
7.1	Pionen data centre.	111
8.1	Online information retrieval technology-in-practice enacted by search engine engineers.	120
11.1	Mr. Head telling stories about the future of communication.	175
11.2	The ‘newsbeagle’.	175
11.3	Estonian virtual embassy exterior.	176
11.4	Estonian virtual embassy interior.	177
11.5	Information board with links.	178
11.6	Advertising the motion picture *The Singing Revolution*.	178
11.7	Estonian virtual embassy exterior.	179

Foreword

This book is the result of a multidisciplinary research initiative at Södertörn University, aiming to integrate research on culture and technology connected to a number of disciplines at the School of Culture and Communication: Philosophy, Media and Communication Studies, Art History, Aesthetics and the study of Practical Knowledge. Since its formation in 2007 a number of interdisciplinary conferences and symposia have been arranged by the initiative, all with the critical focus on the philosophy of technology and related matters, many of which have also resulted in edited volumes. This is also the background for this volume, which seeks to critically examine cultural technologies conceptually and empirically, that is, the technologies by which culture is produced, entertained and reproduced. Many of the chapters were presented in their first versions at a symposium co-arranged by Södertörn University and the Nobel Museum in Stockholm over two days in October 2009. I am deeply indebted to my co-arranger of the event, Rebecka Lettevall, then research director at the museum. I also want to thank those present (many more than could be fitted into this volume) for their engagement and for the comments and stimulating debates at that occasion.

The research initiative has been generously funded by the Baltic Sea Foundation, as are many of the empirical and theoretical projects that have been carried out within its framework (all listed at www.ostersjostiftelsen.se). Specific financial support with language editing has been provided by the publications committee at Södertörn University, for which the committee has my sincere thanks. Judith Rinker provided me and other non-native English speakers with a more nuanced language.

Two chapters have already been published in the course of the editing of this book, and they are republished here with permission from the rights holders. William Uricchio's chapter has been published in 2011 in Visual Studies, 26(1), 25–35. Maria Bakardjieva's and Georgia Gaden's chapter first appeared in Philosophy & Technology (2011, Online First). It is reproduced here with kind permission from Springer Science+Business Media B.V.

Stockholm in October 2011
Göran Bolin

1 Introduction

Cultural Technologies in Cultures of Technology

Göran Bolin

In late modern society, we are constantly surrounded by technologies of various sorts and of an increasingly complex nature. This fact is hardly surprising to anyone. Although the specific quality of the technological penetration might differ depending on where you live, and the sophistication of the gadgets you have access to might vary depending on the living standards of your specific region, our working and private lives are permeated by technologies. The most obvious of these are naturally the material technologies. These are manifest, tangible tools or machines that we surround ourselves with—cars, airplanes, nuclear power plants, washing machines—and when it comes to media and communication technologies, there are computers, screens, mobile phones, television sets and advanced machines for measurements, calculations, estimations, and so on. Although material technologies, or technological hardware if you will, are the most obvious of technologies, there are also other ways we can think of technology. In addition to the technological hardware we have the technological software: the technologies of doing things. These are the practices, the experience-based ways of doing, and the methods used for achieving ends in our professional and amateur worlds. Accordingly, both the technological hardware and software are inscribed in the concept of technology.

The concept of technology has its origin in the Greek combination of technē (τέχνη), which stands for art or craft, and -logía (-λογία), which refers to the study of something. German philosopher Martin Heidegger (1954/1977) famously wrote on the essence of technology, pointing to technē as that which 'reveals', a 'bringing-forth'—and thus also something related to knowledge and 'truth': 'Technology is a mode of revealing. Technology comes to presence [*West*] in the realm where revealing and unconcealment take place, where *alētheia*, truth, happens' (Heidegger 1954/1977: 13; square parenthesis in original).

As pointed out by Swedish philosopher Hans Ruin (2010), this refers back to Aristotle (n.d.), who, in his *Nicomachean Ethics* (Book VI), regarded technē as one of the five intellectual virtues besides 'scientific knowledge, prudence, wisdom, and intelligence' through which man can get in touch with that which is true (as quoted from Ruin 2010: 184). Technē, in Latin translated as *ars*, referred to 'an intellectual capacity which has to do with making or creating (in Greek poiesis) something new in accordance with a reflexive rational capacity. As such it also has to do with truth and falsity' (Ruin 2010: 184).

Technē is thus the activity of gaining insight through cultural or artistic practice. In the writings of Aristotle, technē clearly has a different meaning than what it usually has today, as was Heidegger's point as well, but it might be good to remind ourselves of its Greek roots, not least because of the affinity between technology and knowledge. So, this meaning of the concept from antiquity—the 'Greek conception of technology' as Heidegger called it, according to Ruin—as being the study of arts and crafts, as well as its philosophical connection to truth, has drifted—in English just as in Swedish (and other languages)—and in everyday parlance technology usually refers to the material tools and machines already described, as well as to social and cultural techniques or practices (ways of doing things) employed in various situations. If we think of how we in modern life have developed social techniques for organizing our daily lives at work, we can see that business meetings, kick-offs, symposia, retreats and other practices are developed in order to reach ends, and to make our daily routines more efficient (sometimes even more pleasurable). Historian Lewis Mumford (1967) even considered society a megatechnology: the organizational principle of division of labour, the structured ways social life is formed, altogether form a giant technology for ordering and adjusting our common life together.

CULTURES OF TECHNOLOGY

Reflected on in this way, we can see that our lives truly are permeated by technologies. Not only do we surrounded ourselves with machines and technical aids of different sorts, we also constantly use social technologies to learn how to raise our children, how to arrange our relationships, how to educate the young, how to relax and how to take care of our health. We can then safely say that we live in *cultures of technology*, or technological culture, as Jennifer Slack and Macgregor Wise (2005) label it. These are cultures not only marked by high degrees of technological penetration, but also where technology comes to be celebrated as the solution to most problems.

Already early in the industrialization process, national industrial exhibitions were arranged, beginning in Paris in 1798 with the exhibition on the Champs de Mars (Benjamin 1955/1983: 165). These exhibitions concentrated around the display and demonstration of technical advances within the then slowly emerging industries of the gradually modernizing nation-states. This focus on the technological developments within the industries followed as a theme in the World's Fairs, the first of which was the Great Exhibition at the Crystal Palace in London in 1851. The technological focus was clearly revealed in the catalogue:

> Other nations have devised means for the display and encouragement of their own arts and manufactures; but it has been reserved for England to provide an arena for the exhibition of the industrial triumphs of the whole world. She has offered an hospitable invitation to surrounding nations to bring the choicest products of their industry to her capital, and there to enter into an amicable competition with each other and with herself. (Catalogue text for the World's Fair 1851, cited in Friedberg 1993: 82)

The World's Fairs addressed the 'Industry of All Nations', according to the catalogue (Friedberg 1993: 82), and the symbol of industrial development the catalogue text mentions was equalled to technological development. Furthermore, technological advancement was also popular among visitors, which is also why Benjamin saw them as forerunners to the entertainment industry, which had not yet emerged at the time. A phenomenon such as electric light contributed to the public success of the Chicago Columbian Exposition of 1893, if measured in number of visitors—27.5 million fairgoers paid entry to the exposition to watch, among other spectacles, the Edison Tower of Light (Marvin 1988: 171f). Not least the successively new media technologies became the main attractions at the fairs: In 1855, the technology of photography had a special section of its own; mechanical typesetting was introduced and displayed in London 1862; the telephone in Philadelphia 1876; the phonograph in Paris 1878; wireless in Buffalo 1901; television in Chicago in 1933; and so on (Roche 2000: 160).

The World's Fairs are but one of the arenas where technology is introduced, celebrated, marvelled at, and sometimes also criticized. In cultures of technology, technology also becomes thematized in literature, art, film, music and other cultural expressions. Consider for example the science fiction genre, or the high-tech detective drama, or the spy novel. Naturally, in cultural expression we see both technological utopia and dystopia where the gadgets used by, for example, James Bond in the novels and even more notably in the films, represent the affirmative stance, while the horror movie, especially if mixed with science fiction, often takes the dystopian perspective, representing technology gone berserk. Of course, all these technological self-reflections are indicative of (late) modernity and can be read as symptoms of the changes brought on by the modernization process, and the accompanying need to reflect on the consequences of techno-social progress.

In cultures of technology, then, cultural praxis and expression are centred on technology. Technology becomes the overarching value that sets the agenda for all spheres of society. Technology becomes teleology, the ultimate goal and the unquestioned positive value—and all who question this value are considered technophobic and backwards striving, resisting the progress of technological advancement and development. In one of Swedish techno-sceptic Lasse Svanberg's many books on the discourses surrounding technology, he reported on a small badge he once spotted on the jacket sleeve of an engineer: 'Technology is the answer'—under which a parenthesis was added in smaller type, 'But what was the question?' (Svanberg 1983: 180). The wit of the badge builds on this commonly held and unquestioned assumption of equalling technology with progress and general human development.

Even if most people celebrate, or at least have an unproblematic relation to, technology, there are naturally also some that question technological development, and take a sceptical stance. Techno-sceptics are quite often accused of being Luddites, technophobic standing in the way of progress, referring to the Luddite movement of the early 19th century that resisted technological developments in the textile factories of England through 'negotiating, bargaining, striking, burning, rioting, and machine-breaking' (Slack and Wise 2005: 70). As Kevin Robins and Frank Webster argue, Luddism should be historically contextualized as a response to 'the displacement of traditional social relations by the relations of laissez-faire capitalism' (Robins and Webster 1999: 39), where the changes to social relations were brought on by the mechanization of production and the deskilling of labour. However, despite the historical contextualization made by Robins and Webster, Edward Thompson (1968) and others, Luddism has mainly lived on as a derogative label for those regarded as backwards-striving enemies of technological progress.

And there can be no doubt that technology has had consequences on social relations, in a variety of ways. As Joshua Meyrowitz (1985/1986) argues in his oft-cited book *No Sense of Place*, electronic media have had clear impacts on our social behaviour, exemplified by the changed relations between the genders, between citizens and politicians and between generations. Through the media, most notably television in the case of Meyrowitz (after all, his book is from 1985), men and women gain access to each other's 'back regions' and get glimpses of what those of the opposite sex talk about in single-gendered settings. In a similar way, argues Meyrowitz, citizens gain access to the back regions of politics and politicians through the media, and the young in corresponding ways gain insight into the worlds of older generations in ways not possible before electronic media became an established ingredient of our everyday lives.

While Meyrowitz is thinking about the changed relations between front and back regions, that is, in relation to our private or intimate spheres, we can also see how technology is altering the more semi-private sphere of work. Not only has the increased technological sophistication and the increased technification of vocations made dead end jobs disappear. Today many white-collar jobs disappear by the minute, in favour of the self-administration of employees with the aid of digital and mobile technologies. One such occupation is secretaries, a job that largely has disappeared due to the widespread use of computers, and the increased demands on self-administration (not least within universities) (cf. *Dagens Nyheter*, 19 June 2011, p. E4, where also other examples are given on recently vanished jobs).

However, jobs do not only disappear. Admittedly, there are also new occupations that appear on the job market. Some of these are also unskilled jobs, and they are most often outsourced to low-wage countries. Some of these jobs, such as telephone service and computer support, can take advantage of ICTs, where call centres and support serves customers on the opposite side of the world. One of the more spectacular job niches made possible by the widespread access to computers and high-speed broadband technology is Amazon's Mechanical Turk (mturk.com), were mobile phone apps that build on recognition (of food, for example) can be fed with information from a worldwide source of what you might call 'micro-workers', who can identify colours, meals, and so on, and send this information back to the 'requester', with very small sums in return for their work (sometimes as little as $0.02). So, some of the things we think are the result of cleverly constructed algorithms, programmed to recognize amount of calories in meals, for example, as in 'Meal Snap', are really the result of dispersed global labour (see also mealsnap.com).

CULTURAL TECHNOLOGIES

In cultures of technology, however, culture also has become a tool in and of itself, an arbiter for the achievement of various other ends. It has often been used as a means to reach economic, political, social and other goals. Seen from such a perspective, culture *is* technology: a *cultural technology*. This could be exemplified by the quote from the main architect behind the European Community (EC), Jean Monnet, when he retrospectively lamented, 'If we were beginning the European Community all over again, we should begin with culture' (quoted from Schlesinger 1987: 222). In this sense culture is the tool by which social solidarity, and political unity, is built. Many have taken approaches such as this *ad notam* today, which can be seen in phenomena such as nation branding or destination branding. But culture is also used as a means in advertising, promotion and in the ways in which cultural phenomena such as search engines, social networking sites and the social web generally, are used as means to create economic value. Often culture even becomes reduced to a means, and its existence only justified if it serves external demands. If we analyse the discussions on cultural policy (in Sweden and elsewhere), culture constantly has to be justified in terms of serving other purposes (not least economic; that it creates job markets, that it attracts attention to non-cultural phenomena, etc.). However, even if reduced to the mere means of other kinds of production, the production processes involving cultural technologies has their by-products, where cultural patterns evolve—sometimes beyond the scope of economic, social or political use value, and sometimes in conjunction and combination with these values (see the more elaborate discussion on this in Bolin 2011).

Take nation branding, which is a quite recent phenomenon, and one which I have spent some energy on writing about elsewhere (Bolin 2006a, 2006b, 2010; Bolin and Ståhlberg 2010). Nation branding efforts typically engage cultural technologies in order to construct favourable images of nation-states, in order to raise interest from corporate business, and hopefully get them to invest in the branded country (cf. Dinnie 2008a), or to take advantage of this nation-state's production or consumption force (Bolin and Ståhlberg 2010). Nation branding processes are adopted by many states, but they are arguably as most obvious in post-Soviet and other post-colonial states that have felt a need to rebrand, and cast away their Soviet, French, British or other colonial heritage and its accompanying imageries imposed 'from outside'. Accordingly there have been several such case studies published lately, focussing on nation-states with felt needs to change their appearance on the world stage (e.g., Aronczyk 2008; Dinnie 2008b; Jansen 2008; Kaneva and Popescu 2011; Ooi 2004; Rajagopal 1999).

In nation-branding campaigns, culture is most often used as a prismatic focus, such as when the campaign Brand Estonia was centred on the arrangement of the Eurovision Song Contest (ESC) in 2002, when Estonia as the first post-Soviet state were to arrange the final for this very popular European media event after having won the competition the year before. Through engaging the British PR agency Interbrand, the Estonian government launched a campaign that resulted in 'the deployment of targeted, strategic messages communicated using visual and verbal components', that would supposedly 'enable Estonia to achieve greater success in attracting foreign-direct investment, to expand our tourist base beyond Sweden and Finland, and to broaden European markets for our exports' as quoted from the foreword in the final report of the campaign, the portfolio *Eesti Stiil* (Estonian Style). This portfolio included, besides a description of the background research of the campaign (focus groups, questionnaires, etc., to key target groups), a bank of pictures of people and things that would symbolize Estonia, an Estonian colour palette, a specific typeface, a logotype, and examples of Estonian handicraft among other things (see Bolin 2006b for more details). Since the ESC is broadcast in most European countries, as well as quite a few outside of Europe (e.g., Australia, Japan), this event not only gathers a viewing and voting audience of some 200–300 million viewers. It is also covered intensely by the tabloid and weekly press the weeks before the event in May each year, a consequence that the Brand Estonia campaign took full advantage of through planting strategically targeted messages about Estonia in the hands of the 1,500 accredited journalists present for the occasion (Bolin 2006b).

From the example Brand Estonia we can see that not only were the efforts to attract foreign investment and tempt tourists to travel to Estonia centred around the ESC, one also worked with cultural symbols such as logotypes, typefaces, design and handicraft, all of which became the tools for what in the end are pure economic goals. This is naturally nothing new—all advertising build on these principles, where media texts and cultural expressions are the lures that help construct the audience commodity, a commodity that can ultimately be sold to advertisers. This is also how cultural and aesthetic value are subsumed the overarching goal of producing economic surplus value. This is also the logic that underlies the will of governments to engage in social networking sites such as Facebook, or to take part in virtual worlds extending their activities to the digital domain (eg. Bengtsson 2011, and Bengtsson's chapter in this volume).

HISTORICIZING CULTURAL TECHNOLOGIES

Cultural technologies, however, have not appeared in late modern society from nowhere. Arguably language is our first cultural technology, and one each individual is in possession of. Not all individuals master language to the same degree, however. Some are more rhetorically refined, which means that they have developed linguistic techniques to inform, control or to persuade more effectively.

If language could be said to be our first cultural technology, we can see that language as a communicative device have met with more and more refined technological contemporaries. In fact, many technologies in modern society are media and communication technologies: tools for the dissemination of information, for entertainment and for the organization of increasingly complex societies. If it is true, as John Dewey (1916) once remarked, that society exists in communication, we can safely say that modern society through its increasingly complex ways of communicating, exists in mediated communication (in the sense that we extend our communicative capabilities through communication aids that exceeds the human body). The ways in which communication is technologically mediated of course changes over time, and we are continuously provided with ever-new means of mediated communication.

Culture as lived and experienced in advanced, late capitalism (or late modernity, if you will), then, is increasingly imprinted by media technology. It is produced but also consumed with the help of technology. The means of production used to produce these cultural expressions is also technologically complicated. Since the rise of modern mass media the complexity of the production of culture has indeed increased, beginning with the rotary presses of the mid-19th century, over film production, television, and so on. Today, the means of production have become technologically even more advanced, built around algorithms and computer software in sophisticated networks of communication.

However, we can see that we also surround ourselves more and more with technologies in order to *consume* cultural expressions: In digital cultural landscapes we are increasingly dependent on machines in order to access images, music, films, television and all the multimedia expressions that surround us. Paradoxically, as cultural expressions are increasingly freed from their tangible form due to digitization and can be consumed in a variety of ways, this also means that we become dependent on material *means of consumption* in order to decode cultural texts and make them take on consumable form, allowing them to be interpreted.

With the interactive possibilities of Web 2.0 we are moving into yet another stage of media technologies, whereby access to the means of media production is disseminated to an ever-greater portion of media users (at least in the technologically advanced countries of the world). On the one hand, this means a dissemination of power to express oneself in contemporary culture, to be in control of the cultural technologies. On the other hand, the opportunities provided by the new media technologies also open up for new means of surveillance and control on the part of the media industries, providing for new business models built around Internet traffic, behavioural targeting, and so on (cf. Van Couvering in this volume).

While there are new tools that can be used by the media and culture industries, indeed affecting the ways the industries work (cf. Miller in this volume), there are also changes brought on by technology that concern our uses of new digital media. Continuously more refined ways of representing the world we live in are also arguably affecting our relation to that world, as the new means at our disposal give us new ways to perceive (cf. Uricchio and Rothenbuhler in this volume). At a more fundamental level, these new technologies also affect how we come to learn about the world, effectively changing our epistemological abilities and comprehensions. Knowledge in itself becomes deeply intertwined with technology (cf. Braman in this volume), something that the self-reflexive researcher can also have an impact on and thus take advantage of (cf. Hermes in this volume).

CONCEPTUALIZING CULTURAL TECHNOLOGIES

The concept of cultural technology has been used in different contexts. Most of the time it is used synonymously with 'the media', as a qualification and alignment with the theoretical legacy of thinkers such as Marshall McLuhan's (1964) *Understanding Media: The Extensions of Man* or Raymond Williams's (1974) *Television: Technology and Cultural Form*. Thus, one can read books on cultural technologies like Barbara Gentikow's (2010) *Nye fjernsynserfaringer* (New television experiences), in which cultural technology refers to television and how this specific mass media technology is used and experienced. In this sense all media are per definition cultural technologies, as they are (one of) the tools by which culture is formed, reproduced or changed. In Gentikow's writing the concept is somewhat undertheorized, and she does not take advantage of its conceptual possibilities.

However, other scholars have been more inclined to theorize the concept. For Friedrich Kittler, initiator of the Helmholtz Center for Cultural

Technologies at Humboldt University, Germany, '"media" are first and foremost cultural techniques that allow one to select, store, and produce data and signals', in the words of his colleague at the centre, Sibylle Krämer (2006: 93). The strong structuralistic element in Kittler's (e.g., 1986/1999) conceptualization, and the transmission perspective privileged, give it a determinist bias that fits less well with the intentions in this book. To Kittler, subjectivity is seemingly an effect of the media as cultural technologies, which means that there is little room for individual action and experience in relation to them. Kittler 'gives us a media studies without people', as John Durham Peters (2010: 5) concludes in his foreword to Kittler's (2002/2010) *Optical Media*. This means that there is also very little room for the interrelation between culture and technology.

Canadian media scholar Jody Berland, working in the tradition of McLuhan although with a more coherent theoretical approach that also includes theories of the production of space in the following of Henri Lefebvre (1974/1991), might be the one who has most often returned to the concept of cultural technologies. In her recent collection of essays, *North of Empire* (2010), summarizing her work over the past two decades, Berland describes cultural technologies as integrated relations between technology, space and power; they are, on the one hand, 'spatializing mediators of power', and on the other, 'assemblages of space, technology, identity and meaning' (Berland 2010: 303). This conceptualization, which encompasses both the structuring aspect of culture and technology as well as the assemblages (the structured results of these processes), is a more dynamic and appealing way to theorize cultural technologies. This specific dialectic quality of the concept shares Bourdieu's (1980/1990: 52ff) concept of 'structuring structures' and 'structured structures', and allows for a more nuanced analysis of cultural and technological phenomena. The structures we have to adjust to in our personal and professional lives are not static, but are themselves structured by various cultural and societal forces, and we need to acknowledge that this is dynamic.

The way the concept is fitted into this volume through the represented perspectives on culture and technology is more in line with how Berland uses it, and has less to do with the technological determinist usage of Kittler. It is used to highlight the technological side of culture, how culture is formed historically as a process or set of technological practices but also how it can be appropriated as a tool in a social setting where technology has become one of the core values of society.

To summarize, one could say that the concept of cultural technologies as used here has four dimensions. First, and in its most general sense, it refers to technologies of communication—speech, writing and the technical aids with which we communicate and represent the world around us. These are cultural technologies that help produce cultural expressions in art and popular culture. We could call this the *expressive dimension* of cultural technologies. Second, it refers to those practices and ways of organizing our lives that we adopt in private or professional contexts; how we as parents go about bringing up our children, making them competent cultural and social beings, but also those rules and practices by which we organize our professional lives, and how these practices result in cultural patterns. We can call this the *social dimension* of cultural technologies. Third, such organizational principles can also include the more physical organization of professional lives, for example, in the buildings of the media, the architectonics of media houses (cf. Ericson and Riegert 2009 and Colomina 1994) and how these buildings are constructed in order to foster specific work ethics or promote efficiency. This is the *physical dimension* of cultural technologies. Fourth, we have the dimension where culture is subordinated technology—where culture is not the end result but the means deployed for other ends. This is the *administrative dimension* of cultural technologies.

Heidegger thought of technology as a 'bringing-forth', something that produces something. Interestingly enough, the Latin combination of *pro* and *ducere* in the word production also refers to something that can be called 'bringing forth' (pro=forward and ducere=to lead).[1] Technology, then, is deeply connected to the production of knowledge, and hence to epistemology. This is not only because the more administrative pedagogical technologies used in teaching and learning bring with them tools for efficiency in educational spheres. It is also connected to the critique of technology and its critical uses, that is, the deeper penetration of technological processes that help us be aware of the worlds (in the phenomenological sense) we inhabit, be it in the observation of new technologies of seeing (Winston 1996, 1998) or those of hearing (Sterne 2003).

THE SHAPING OF CULTURE IN MEDIA AND SOCIETY

The collection of essays in this volume discusses both the culture *of* technology that we live in today, and culture *as* technology. Cultures of technology and cultural technologies are discussed in the chapters of the book, focussing on a variety of examples from different national contexts, and from historical as well as contemporary perspectives. It contains analyses of technological phenomena as well as epistemological discussions on the uses of technology. The edited collection is structured in one introduction and three sections, focussing first on the historical dimension, second on the epistemological nature of technologies, and third on the uses of cultural technologies.

In the first section on *Histories of Cultural Technologies*, in Chapter 2 William Uricchio discusses changes in visual perception as a consequence of new representational technologies and 'algorithmic interventions' in the subject–object relationship provided by visual technologies such as *Photosynth*. As Uricchio argues, this is not merely a change from the uniqueness of the subject–object relationship of artists of the modern era to the multiple people's perspective that we might think technologies like Photosynth provide us with. It is also something deeply embedded in technology that reflects the 'interplay of algorithmically determined points of uniqueness' that is activated by the navigational work of the user within these technological frameworks.

In Chapter 3, Eric Rothenbuhler historicizes another perceptual transformation related to technology, although his focus is on the audible rather than the visible. His discussion is centred on the changes in fidelity that occurred with the change from analogue to digital technology, and how accessibility and convenience have become more important than sound quality in contemporary music cultures. Up until the introduction of the CD, Rothenbuhler argues, music production had strived towards increasingly better sound quality. Since digitization we have seen a decrease in quality to the benefit of greater accessibility to recorded music, whereby today we can access most music over the web, in compressed formats and with less dynamic sound range. The consequence of this change, it is argued, is a shift from the strive for audible fidelity, truth and perfection that was indicative of the analogue sound culture to a strive for efficiency, convenience and price, effectively transforming the listener into a consumer, driven by market logics.

In Chapter 4, Mats Björkin continues the discussion of the changes brought on by digitization, in his case connected to the distribution techniques that have been developed for media content through tracking systems and BitTorrent technology. The question Björkin focuses on is how the cultural object—primarily films and video clips—is transformed due to these changes, especially through file-sharing techniques. As Björkin argues, new skills are required for those who seek to orient themselves among the vast amount of content available.

The distribution of media content is highly dependent on infrastructures for the dissemination of information and communication. As part of such infrastructures, the satellite dish can be said to symbolize transnational communication flows. In Chapter 5, Lisa Parks discusses the ways infrastructure becomes intelligible to media users, and how the 'intelligibility of infrastructures is interwoven with the politics of knowledge, place and aesthetics'. Infrastructure is often concealed and not openly accessible to citizens, but the satellite dish has a manifest materiality around which discourses of taste, ethnicity and social status are formed. Parks shows how the satellite dish, through this material quality, can be a 'platform for thinking through key issues of technological literacy, social power, and resource use'.

In the second section of the book—*Epistemologies of Cultural Technologies*—the relation between technology and knowledge is in focus. These are chapters that broadly thematize the ways cultural technologies help corporate actors as well as policy makers and individual citizens gain knowledge about the world. In Chapter 6, Toby Miller discusses the consequences of new surveillance techniques, and how the market logics actively embrace active audiences who share their tastes and preferences. Even the acts of resistance to hegemonic ideologies become just another market niche in the eyes of the television industries, Miller argues, pointing to how 'our viewing and social identities are governed and commodified and our mental labor sold to others without our consent or remuneration'.

The intelligence produced through surveillance technologies is deeply dependent on computational power and storage capacity. The data centres and server farms that serve Google, Facebook, Twitter and other services that build on the ability to process enormous quantities of digital information and data are of fundamental importance for the digital infrastructures that enable cloud computing, data mining and harvesting (and surveillance). In a case study in Chapter 7, Peter Jakobsson and Fredrik Stiernstedt analyse discourses around this specific centralization of information, centred on 'three different but related narratives that operate on three time scales: geological, historical and technological time'. Underlying these three temporal discourses, they also find a deeper dialectic of durability and ephemerality that they consider inherent in discourses on digital data and the information society: on the one hand, the ephemeral quality of digital information, and on the other, the tendency of data centres to occupy the same societal space as the library, symbolizing cultural continuity.

One of the main activities on the Internet is web searches. Google, Yahoo!, Bing and other search engines compete for the domination of the search market, because the larger the information base, the better the search behaviour commodity can be refined. And around the business of search, a range of new occupations has arisen, web optimizers being one of them. In her chapter, Elizabeth Van Couvering analyses how 'search engine bias comes to be embedded into both local practices and larger systems', that is, how certain web sites are becoming privileged in the search system over other sites. She then goes on to show how this indicates a conflict between 'the search engine as a private media enterprise and the search engine as a public good'.

The intelligence produced by search engines is but one of the sources of knowledge in the digital environments. In Chapter 9, Sandra Braman takes on a wider discussion of the epistemological implications of technology, especially related to information policy. She suggests that information policy actually equals epistemology policy, because communicating social, aesthetic or other facts also means producing knowledge of the world around us. Taking her departure in John Locke's four elements of the fact—the perceptual entity, the subjective experience, the symbolic expression to others and the group discussions in which we construct reality in social interaction—she asks the fundamental question for epistemological inquiry: 'What knowledge can we produce about the decisions we make about how to produce knowledge?'

The third section of the book covers the *Uses of Cultural Technologies*. In this section, three chapters analyse cultural technologies as means to other ends. In Chapter 10, Maria Bakardjieva and Georgia Gaden discuss the uses of Web 2.0 technologies for the construction of self. Following Foucault's famous 'technologies of the self', Bakardjieva and Gaden reflect on the abilities of blogs and social networking sites to function as cultural technologies in the service of the individual in his or her self-construction, not least through their abilities to provoke reflexivity and play.

Whereas Bakardjieva and Gaden focus on the individual construction of self, in Chapter 11 Stina Bengtsson focuses on the formation of collective social identities. Her focus is on the Estonian embassy in the virtual world Second Life, and how the Estonian government used this forum as a cultural technology and 'offered a national space *in* the media' where citizens could take part in events, get Estonian news, and so on. Contrary to other embassies launched in Second Life, and despite the fact that it was launched by the Ministry for Foreign Affairs, the Estonian embassy gradually changed its address from the usual external address to foreign investors and tourists (the mark of nation-branding campaigns) to a more domestic address, directed at Estonian citizens—making it a tool for nation building rather than nation branding.

In the final chapter, Joke Hermes reflects on the possibilities of technology to serve qualitative, empirical audience research. Hermes argues that 'technology needs to be rescued from its ambivalent status' in the humanities, and, on the one hand, should be acknowledged as an area of study in itself, and on the other, also employed as an aid in interpretive approaches that seek to analyse meaning-making processes, for example, in qualitative audience research. Comparing technological aids of the social sciences and the humanities, she concludes that while there have been dramatic methodological developments within the social sciences, making the means of statistical calculation accessible to a wider circle of scholars, the technological instruments for conducting qualitative research have not followed pace. She especially wants to see analytical software developed, not only to analyse large qualitative datasets but also to guide longitudinal consistency. In developing such software, she argues, we need to see new types of cross-disciplinary collaborations between qualitative researchers and software developers.

Taken together, these chapters are supposed to give an insight into the varieties of our cultures of technology, indicating the wide spectrum of media and other technologies that in conjunction with each other form this culture: expressively, socially and physically. In the chapters are also given some examples on cultural technologies, the cultural tools by which we administer or construct our selves and our environments individually, socially and professionally.

NOTES

1. For a fuller account of the Latin root of production, see the Oxford English Dictionary. Available at http://www.oed.com/view/Entry/151994?redirectedFrom=production#eid. Last accessed 2011–10–05.

REFERENCES

Aristotle. n.d. *Nicomachean Ethics*. Translated by W. D. Ross. The Internet Classics Archive. Available online at http://classics.mit.edu//Aristotle/nicomachaen.html. Last accessed 2011–09–27.

Aronczyk, M. 2008. Living the Brand: Nationality, Globality and the Identity Strategies of Branding Consultants. *International Journal of Communication*, 2, 41–65.

Bengtsson, S. 2011. Virtual Nation Branding: The Swedish Embassy in Second Life. *Journal of Virtual Worlds Research*, 4(2). Retrieved from https://journals.tdl.org/jvwr/article/view/2111. Last accessed 2012–03–08.

Benjamin, W. 1955/1983. Paris—the Capital of the Nineteenth Century. In *Charles Baudelaire: A Lyric Poet in the Era of High Capitalism*. London: Verso, 155–176.

Berland, J. 1992. Angels Dancing: Cultural Technologies and the Production of Space. In *Cultural Studies*, edited by L. Grossberg, C. Nelson and P. Treichler. New York: Routledge, 38–55.

Berland, J. 2000. Cultural Technologies and the 'Evolution' of Technological Cultures. In *The World Wide Web and Contemporary Cultural Theory*, edited by A. Herman and T. Swiss. London: Routledge, 235–258.

Berland, J. 2009. *North of Empire: Essays on the Cultural Technologies of Space*. Durham, NC: Duke University Press.

Bolin, G. 2006a. Electronic Geographies: Media Landscapes as Technological and Symbolic Environments. In *Geographies of Communication: The Spatial Turn in Media Studies*, edited by J. Falkheimer and A. Jansson. Göteborg: Nordicom, 67–86.

Bolin, G. 2006b. Visions of Europe: Cultural Technologies of Nation-States. *International Journal of Cultural Studies*, 9(2), 189–206.

Bolin, G. 2010. Media Events, Eurovision and Societal Centers. In *Media Events in a Global Age*, edited by N. Couldry, A. Hepp and F. Krotz. London: Routledge, 124–138.

Bolin, G. 2011. *Value and the Media. Cultural Production and Consumption in Digital Markets*. Aldershot: Ashgate.

Bolin, G. and P. Ståhlberg. 2010. Between Community and Commodity: Nationalism and Nation Branding. In *Communicating the Nation: National Topographies of Global Media Landscapes*, edited by A. Roosvall and I. Solivara Moring. Göteborg: Nordicom, 79–101.

Bourdieu, P. 1980/1990. *The Logic of Practice*. Cambridge: Polity Press.

Colomina, B. 1994. *Privacy and Publicity: Modern Architecture as Mass Media*. Cambridge, MA: MIT Press.

Dewey, J. 1916. *Democracy and Education*. New York: Macmillan.

Dinnie, K. 2008a. *Nation Branding: Concepts, Issues, Practice*. Oxford: Butterworth-Heinemann.

Dinnie, K. 2008b. Nation Branding and Russia: Prospects and Pitfalls. *Russian Journal of Communication*, 1(2), 199–201.

Ericson, S. and K. Riegert. Eds. 2010. *Media Houses: Architecture, Media, and the Production of Centrality*. New York: Peter Lang.

Friedberg, A. 1993. *Window Shopping: Cinema and the Postmodern*. Berkeley: University of California Press.

Gentikow, B. 2010. *Nye fjernsynserfaringer. Teknologi, bruksteknikker, hverdagsliv*. Oslo: Højskoleforlaget.

Heidegger, M. 1954/1977. *The Question Concerning Technology, and Other Essays*. New York: Garland Publishers.

Jansen, S. C. 2008. Designer Nations: Neo-Liberal Nation Branding—Brand Estonia. *Social Identities*, 14(1), 121–142.

Kaneva, N. and D. Popescu. 2011. National Identity Lite: Nation Branding in Post-Communist Bulgaria and Romania. *International Journal of Cultural Studies*, 14(2), 191–207.

Kittler, F. A. 1986/1999. *Gramophone, Film, Typewriter*. Stanford, CA: Stanford University Press.

Kittler, F. A. 2002/2010. *Optical Media: Berlin Lectures 1999*. Cambridge: Polity Press.

Krämer, S. 2006. The Cultural Techniques of Time-Axis Manipulation: On Friedrich Kittler's Conception of Media. *Theory, Culture & Society*, 23(7–8), 93–109.

Lefebvre, H. 1974/1991. *The Production of Space*. Oxford: Blackwell.

Marvin, C. 1988. *When Old Technologies Were New: Thinking About Electric Communication in the Late Nineteenth Century*. Oxford: Oxford University Press.

McLuhan, M. 1964. *Understanding Media: The Extensions of Man*. New York: McGraw-Hill.

Meyrowitz, J. 1985/1986. *No Sense of Place: The Impact of Electronic Media on Social Behaviour*. Oxford: Oxford University Press.

Mumford, L. 1967. *The Myth of the Machine: Technics and Human Development*. Vol. 1. New York: Harcourt, Brace, Jovanovich.

Ooi, C.S. 2004. Brand Singapore: The Hub of 'New Asia'. In *Destination Branding: Creating the Unique Destination Proposition* (2nd ed.), edited by N. Morgan, A. Pritchard and R. Pride. Burlington, MA: Butterworth-Heinemann, 242–260.

Peters, J. D. 2010. Introduction: Friedrich Kittler's Light Shows. In F. A. Kittler: *Optical Media: Berlin Lectures 1999*. Cambridge: Polity Press, 1–17.

Rajagopal, A. 1999. Thinking Through Emerging Markets: Brand Logics and Cultural Forms of Political Society in India. *Social Text*, 17(3), 131–149.

Robins, K. and F. Webster. 1999. *Times of the Technoculture: From the Information Society to the Virtual Life*. London: Routledge.

Roche, M. 2000. *Mega-Events and Modernity: Olympics and Expos in the Growth of Global Culture*. London: Routledge.

Ruin, H. 2010. Ge-stell: Enframing as the Essence of Technology. In *Martin Heidegger: Key Concepts*, edited by B. W. Davies. Montreal: McGill-Queens University Press, 183–194.

Schlesinger, P. 1987. On National Identity: Some Conceptions and Misconceptions. *Social Science Information*, 26(2), 219–264.

Silverstone, R. and E. Hirsch. Eds. 1992/1994. *Consuming Technologies: Media and Information in Domestic Spaces*. London: Routledge.

Slack, J. D. and J. M. Wise. 2005. *Culture + Technology: A Primer*. New York: Peter Lang.

Sterne, J. 2003. *The Audible Past: Cultural Origins of Sound Reproduction*. Durham, NC: Duke University Press.

Svanberg, L. 1983. *Bekvämlighetens triumf*. Stockholm: Prisma.

Thompson, E. 1968. *The Making of the English Working Class*. Harmondsworth: Penguin.

Williams, R. 1974/1979. *Television: Technology and Cultural Form*. London: Fontana.

Winston, B. 1996. *Technologies of Seeing: Photography, Cinematography and Television*. London: BFI.

Winston, B. 1998. *Media Technology and Society: A History: From the Telegraph to the Internet*. London: Routledge.

Part I

Histories of Cultural Technologies

2 The Algorithmic Turn

Photosynth, Augmented Reality and the Changing Implications of the Image

William Uricchio

Rien n'est plus ennuyeux qu'un paysage anonyme. (Prosper Merimée)[1]

TERMS OF ENGAGEMENT

I've long been haunted by Jonathan Culler's *Semiotics of Tourism*, with its careful parsing of travel vs. tourism, its astute characterisation of behaviours such as our reflexive denigration of those more tourist-like than we, and its charting of the elusive search for authenticity (Culler 1990). Culler traces the process by which cultural attractions are marked as signs produced by an international system of signification, and responsive to and inscribed within an economic order. Yet he avoids a blanket indictment that would read this process as the mere flattening of the authentic into caricature, complicit with the demands of multinational capitalism (a charge he locates with the 'sentimental nostalgia for the organic'). His essay instead calls for an exploration of the persistent and ubiquitous semiotic mechanisms central to any social order, a task that this essay takes up. It is my contention that we can see evidence of change in the dominant 'semiotic mechanisms', evidence that perhaps speaks to a deeper transformation of the social order. The goal of this chapter is to explore one aspect of that change: the algorithmic construction of the image. While I will not take up the grander challenge of discussing the transformation of the social order, I will address what I take to be one of its key symptoms, namely, cracks in the facade of the subject–object relationship characteristic of the modern era. Let me make clear at the outset that cracks do not a transformation make, but they give us an early warning and an important place to look for further signs of change.

My argument in a nutshell is that over the past decade or so, we have had increased access to new ways of representing and seeing the world, ways dependent on algorithmic interventions between the viewing subject and the object viewed. This intervention has many manifestations, from the changed model of authorship and expertise that Wikipedia represents over and against the Enlightenment paradigm represented by Diderot's encyclopaedia, to the dynamic and location-aware cartographic systems that we can find on our iPhones and TomToms over and against the fixed cylindrical projection grid of Gerardus Mercator's 16th-century maps. In the case of the image, *grosso modo*, the long regime of three-point perspective and its reification of an underlying understanding of subject–object relations constitute the representational order that is exhibiting fractures. And those fractures—the topic of this essay—appear in the form of algorithmic visualisation systems such as Microsoft's Photosynth and image recognition-based augmented reality applications.

In order better to locate the terrain and implications of this argument, we might briefly look at the larger regime in which the algorithmic represents an incursion. Its underlying premise can be found in Culler's assertion of the centrality of semiotic mechanics to the social order, a view that echoes a position developed by Heidegger in his discussion of the *Welt-bild*, where the world as picture 'does not mean a picture of the world but the world conceived and grasped as picture' (Heidegger 1938/1977: 129). Heidegger goes on to specify that the world picture 'does not change from an earlier medieval one into a modern one, but rather the fact that the world becomes picture at all is what distinguishes the essence of the modern age' (Heidegger 1938/1977: 130). For Heidegger, the moment at which the world becomes picture is the same moment that the human emerges as the subject in a characteristically modern subject–object relationship. It marks the birth of the modern. The modern subject–object relationship takes many forms, but can most clearly be seen in our notion of science. Heidegger says,

> We first arrive at science as research when and only when truth has been transformed into the certainty of representation. What it is to be is for the first time defined as the objectiveness of representing, and truth is first defined as the certainty of representing, in the metaphysics of Descartes. (Heidegger 1938/1977: 127)

Heidegger here asserts the fundamental linkage of mode of representation and epistemic (and social) order. The Cartesian metaphysics to which he refers offer a widely accepted definition for the modern project that remains dominant today, even if under siege. In this order, precision and exactitude as embodied in classical mathematical physical science are essential ingredients. Heidegger states, 'Here all events, if they are to enter at all into representation as events of nature, must be defined beforehand as spatiotemporal magnitudes of motion. Such defining is accomplished through measuring, with the help of number and calculation' (Heidegger 138/177: 119).

Culler's general proposition that a social order has central 'persistent and ubiquitous semiotic mechanisms' is embodied by Heidegger's essay, which argues that the modern social order can be defined through a representational system characterised by precisely defined subject–object relations (the world as picture), a metaphysics of exactitude, and an underlying spatiotemporal grid. Descartes himself emblematises this order, thanks to his development of analytical geometry and the Cartesian coordinate system that followed (enabling geometric shapes to be described in algebraic form). He stands for a notion of the modern typified by classical cartography, three-point perspective and the amplified point of view manifest in authorship and the cult of expertise (Diderot). The broad space between Descartes and Heidegger, despite countless historical undulations and discoveries, turns on a consistent logic, on something we might term the *algorismic*, that is, a notion of mathematics as a language of precision, of calculability, of predictability. And it remains deeply familiar to us, pervading our lives whether through our financial and insurance systems, our notions of science or the construction of our default technologies of visual representation. While three-point perspective and clearly defined subject–object relations are not inherently algorismic, their operations are consistent with it and depend upon its logics.

The algorismic that Descartes did so much to promote as a mathematical system, an epistemic regime and even a social order, and that Heidegger and others argued as a defining characteristic of the modern era, shares etymological roots with the term *algorithmic*, but the two have very different definitions and histories. In contrast to the precision, calculability and specificity of the algorismic, an algorithm refers to a process, a program with clearly defined limits, a finite instruction sequence. One is a calculable sum, whose value lies in the correctness of its result; the other a (finite) process, a formula capable of accommodating different values and yielding different results. Each term harkens back linguistically to the great 9th-century Persian mathematician Al-Khwarizmi, whose work appeared in the West by the 13th century, and as mathematical models each can be traced back to the ancient Greeks. Yet their history differs considerably. The algorismic predominated, enjoying a particular boost in use in the 16th and 17th centuries. By contrast, although Euclid demonstrated their operation ca 300 BC, algorithms remained a marginal form of calculation until the second half of the 20th century, when the introduction of computers enabled the efficient processing of programs.2 The mathematical advances of the 20th century owe less to the arithmetic of the algorismic, than to the accelerated development of algorithmic instruction sequences, a process that continues to grow exponentially thanks to the ongoing validity of Moore's Law. The chips in our phones, cameras and bank cards, or the supermarket 'loyalty tags' scanned as we check out, or the cookies and behaviour tracking that companies such as Amazon and Google use to predict and assess our interests, all attest to the pervasiveness of an infrastructure designed to generate data whose meanings can only be deciphered algorithmically. And it is this algorithmic layer that stands between the calculating subject and the object calculated, and refracts the subject-centred world charted by Descartes, that merits closer inspection. Its implications, if we take thinkers like Heidegger and Culler seriously, can be profound.

MARKING VALUE

One of the semiotic strategies that Culler discusses is the 'marker'. Drawing on Dean MacCannell's *The Tourist*, Culler defines a marker as any kind of information or representation

> that constitutes a sight as a sight: by giving information about it, representing it, making it recognizable. Some are 'on-site' markers, such as plaques telling that 'George Washington slept here' or that this vial of dust comes from the moon. Some are mobile markers, such as pamphlets and brochures designed to draw people to the site, give information at the site, and serve as souvenirs or representations off the site . . . The proliferation of markers frames something as a sight for tourists. The existence of reproductions is what makes something an original, authentic, the real thing—the original of which the souvenirs, postcards, statues etc. are reproductions—and by surrounding ourselves with reproductions we represent to ourselves, as MacCannell astutely suggests, the possibility of authentic experiences in other times and in other places. (Culler 1990: 5)

Culler's essay and MacCannell's insights go a long way towards explaining how value is created by collapsing temporalities, overlaying a site or object in the present with a reference to its provenance or past, or even evoking the authentic through the act of reproduction. A crumbling bit of rock in one's hand transforms when we learn that it is a piece of one of Giza's pyramids or a fossilised piece of Diplodocus bone. The process of generating value through the assignment of meaning is a familiar if complicated one, one with its own spatialities and temporalities: The artefacts within the retail outlet, the antique shop and the second-hand store are distinguished by these dynamics rather than by any intrinsic worth. Mere purchase and ownership, no matter how brief, can recast an object from retail to second-hand; and today's second hand in one market could possibly be revalued as a valuable antique in another.

The image, however, has an even more multivalent status. As an artefact, say, in the form of a photograph, the image is collectable, subject to the same ebbs and flows in valuation as other things, particularly those privileging symbolic and cultural attributes. Tastes, both professional (art markets) and vernacular (eBay), help to set these values. And, like souvenirs, the image has the ability to link us back to the people, places and events it represents. Its very presence attests to 'the possibility of authentic experiences in other times and in other places'.

But as Culler suggests, the image can also be complicit with the valuation of its referent, the place or thing represented. In this case, the relationship between the image and that which it represents can be discursive, amounting to little more than a claim; or it can work through resemblance, something that we simply recognise or take as recognising. The space between the two is slippery and sometimes fraught. Just as the claim that a piece of stone is from Cheops' pyramid changes the stone's status, so too the (discursive) claim that a photo is of a particular place constructs its (representational) meaning. Claims aren't always accurate; film and television directors routinely move between these two modes of assigning meaning, establishing a location through resemblance—iconic shots of a particular place such as Manhattan—only to cut to shots taken in less expensive locations such as Toronto or Pittsburgh as if the same location. This technique of using discursive strategies to evoke representational assumptions was explored by early Soviet filmmaker Lev Kuleshov, who argued that at least in the sequential image flow of film, meaning is constructed by the relationship of one shot to the next rather than fully residing within any one image.

The assignment—and destabilisation—of meaning can occur within the realm of the image as Kuleshov and his followers argued; and theorists such as Culler have extended the argument, turning it on a different sort of framing, but nevertheless undercutting the assumed stability of imagistic representation and meaning by looking outside it, at the broader signifying context. And more recently, theorists such as Tim Ingold and Sarah Pink (2009) have sharpened the critique by addressing factors such as multisensory settings, movement and place, reminding us that there is more to the image than meets the eye. The idea of the image as 'becoming' through movement—movement of the world, movement of the (photographing) subject, and movement even of the subsequent viewer is taken up and developed by Pink, and will resonate with the algorithmic applications to which we will shortly turn.

The paradox of the image as a site of meaning and value turns on its reciprocal relationship to the object or place photographed. As discussed, sometimes the location can give the image its significance, and sometimes, as Culler notes, the image can give the location its meaning. Functioning somewhat differently than a replica—say, a miniature Eiffel Tower, which can be seen from any angle and still maintain the legibility of its link to the real thing—the image carries with it a more constrained sense of representation. An extreme close up of one of the tower's beams would not necessarily convey the meaning 'Eiffel Tower'; it could only do this by relying exclusively on discursive claims. Moreover, even at the representational level, where what we would see is a beam or metal surface, assuming that either was recognisable, we would still have to rely on a discursive claim to distinguish *this* beam or surface from any other metal beams or similarly surfaced objects. And such an image, no matter how much one proclaimed its indexical status, would have a difficult time conveying value back to the object. At least when its representational capacities are being called upon (as distinct from its cultural value as art), the value and meaning of a photographic image are constrained by the viewer's ability to recognise and make visual correlations between the two-dimensional image and the thing represented. And in this particular game, convention and point of view matter, strategies that have been finely honed and codified in the course of the modern era, and that are part and parcel of a particular way of seeing and being in the world. Photosynth and certain augmented reality applications offer some intriguing variations on this process, and in so doing, raise some fundamental questions about our visual regime.

MOVEMENT WITHIN THE IMAGE: PHOTOSYNTH

Photosynth (www.photosynth.net), a software application developed in part by Noah Snavley (PhotoTourism) at the University of Washington and acquired and further developed by Microsoft, essentially analyses digital images for sites of uniqueness, generates a three-dimensional point cloud of the represented space and reassembles the images into a near-seamless composite. Viewers can then explore the assembled three-dimensional photographic space in almost any direction (including in depth), with Seadragon technology enabling instant high-definition access to whatever portion of the whole is on the screen. This both precludes any feeling of zooming into an existing picture (a situation in which granularity increases and the authoritative status of the original composition is reinforced) and facilitates the transition from one image space to another. The overlays and transitions from one image to the next are discernable but not obtrusive, and as a viewing option, the contours of the original photos can be marked with white lines ('quads') for navigational purposes (viewers can also see all of the individual photos laid out side by side if they'd like, as well as different angles of the point cloud). Users can build their own 'synths' by crawling the web or particular sites such as Flickr, and tagging images with a particular location ('Piazza San Marco in Venice') or event ('Obama's inauguration'). Synths can be constructed with between dozens and several hundred photos; for the moment, Microsoft recommends staying below a 600 photo limit since the strain on the system's processing power visibly interferes with the experience. Available to the public since 2008, Photosynth's capacities seem to be developing steadily, with major improvements in the synthing of different images introduced in 2010.

In the Photosynth website's words,

> Photosynth takes your photos, mashes them together and recreates a 3D scene out of them that anyone can view and move around in. Different than static photos and video, Photosynth allows you to explore details of places, objects, and events unlike any other media. You can't stop video, move around and zoom in to check out the smallest details, but with Photosynth you can. And you can't look at a photo gallery and immediately see the spatial relation between the photos, but with Photosynth you can.

Very much a work in progress, Photosynth's status as a medium also seems to be in formation, and this description suggests its sites of slippage relative to existing media forms. It manages to wed place and movement through its 'point cloud'—a three-dimensional configuration of sites of visual uniqueness upon which the images are arranged and through which movement is facilitated—a logic with implications that might best be approached by considering Photosynth's limit cases. The system is predicated upon the visual contours of place. Like a jigsaw puzzle, it works from known or given points, and extends outwards, building an ever changing whole by aligning the points of different images and suturing them together. Unlike a puzzle, however, Photosynth addresses a third dimension, and therefore must confront the issue of significant overlap among its images. Since the synths are composited from digital photographs, some of their challenges regard such formative elements as image contrast ratio, composition (angle, proximity, focus) and lens characteristics, all of which are capable of rendering points of uniqueness in ways that the system cannot recognise. If an extreme close-up of several mosaic pieces from San Marco's basilica is tagged and fed into a synth, it may or may not be accepted, depending upon the certainty of its location as extrapolated through the point cloud. If its pattern is distinctive, allowing it to fit uniquely within other shots of the same area, then it will probably be accepted, allowing the synth's user to traverse the plaza even to the point of a close up of the mosaics. Shots that, because of extreme angles or distorting lenses or lighting configurations, are unrecognised are simply not successfully synthed. These tolerances express the operations of the underlying algorithm, reflect the current state of the navigational interface and attest to particular conventions regarding imagistic correctness—and they seem quite dynamic since the capacities of the system grow ever more refined.

As might be expected, Photosynth does particularly well with architectural environments: Stability is a plus and facilitates the construction of point clouds. However, it also yields impressive results in nature. Rather more dynamic than the relative stasis of architecture, nature raises the larger issue of duration. What happens to the points of uniqueness in a forest when blowing wind, or plant growth, or seasonal changes alter the configuration of leaves and branches? One can find quite impressive synths of forests on Photosynth's site, but most share tranquil conditions, a small temporal window, and presumably make good use of this relative stasis. But we might put the question to architectural spaces as well. Mixing images taken during the day and at night will result in a high percentage of rejected images, but what about shots taken before and after, say, a war or earthquake? New chips on a surface or distorted corners of buildings, windows or bricks complicate the construction of a point cloud, exposing the system's durational tolerances and rendering its embrace of historicity as at best ambivalent.

At the other end of the spectrum, what about the 'transient, the fleeting, the contingent' that Baudelaire ascribes to the modern? The most widely seen example of a Photosynth application that addresses this problem is CNN's appropriately titled 'The Moment', a synth of 628 user-submitted photos of the moment that President Obama took his oath of office. Although many of the photos are of the inauguration itself, large portions of the synthed image include crowds of onlookers, dignitaries and even musicians. 'The moment' actually reveals quite a bit of temporal slippage—changes in bodily position, different configurations of the flags in the wind, and so on. The constants, the point clouds, seem to be grounded in architectural detail and the configuration of the podium more than anything else, and the more one examines close shots of the crowd, the more one is reminded of early 19th-century photography, where long-time exposures often resulted in ghost figures. These fugitive images emerged from the gaze locked within three-point perspective and subject to the limits of early photochemical emulsions. With Photosynth, a different mechanism is at play: the tolerances of algorithmic reassembly; but the ephemeral nevertheless seem to pose a very real challenge to the system.

Photosynth permits one to 'wander' through the depicted space, moving deeper into (or farther back from) the frame and exploring off-screen space, whether to the left or right, or the top or bottom of the initial frame. Indeed, the frame (of the computer screen) is rendered arbitrary, like the lens of a magnifying glass in relationship to a newspaper or the viewfinder of a telescope to the night sky. As the Photosynth advertisement notes, the sense of mobility and user-generated interactivity within the image space challenges our inherited definitional assumptions of photography. But it does something more, restoring an experience of movement to the image and the depiction process, vividly illustrating the arbitrary constraints of any one image within a larger reality. Much like Google Street Views (also an algorithmically enabled application), Photosynth assembles points and enables movement within and among them, offering us different ways of

Figure 2.1 Piazza San Marco with the Basilica, by Canaletto, 1730. Fogg Art Museum, Cambridge, MA.

displaying and engaging with visual information. But beyond Photosynth's ability to do this dynamically and in three dimensions, the application distinguishes itself through its subverting of any particular point of view, in the process disrupting the modern configuration of the subject–object relationship, as well as its reliance on collaborative authorship. Let's take these factors in turn.

Consider Canaletto's painting of Piazza San Marco and the Basilica (1730). An impressive display of three-point perspective, the basilica is at the centre of the frame, and the lines provided by the Procuratie Vecchio and Procuratie Nuovo as well as the gutters in the plaza literally map out Canaletto's sight lines. The representation is completely stable, the geometries a testament to the algorismic regime of fixity and precision. There is one and only one position from which the image's geometries work, and the viewer and artist share that point of view. Photosynth, by contrast, draws on some 600 photographs taken from different positions, angles, distances, and with an array of lenses. The user can navigate within and among these, in the process discovering many well-composed images and new ways of seeing the plaza and its buildings. But there is no correct or authorised viewing position, no 'master shot' within which everything else is a recomposition. Instead, there is simply a three-dimensional space made up of many textures and granularities and the means to move within it. Whereas in Canaletto's case, the subject–object relationship is geometrically fixed, in Photosynth's dynamic assemblage the relationship is unstable, subject to the whims of the user and the capacities of the algorithmic. We can move within and across particular points of view, confront radical shifts in angle and inhabit a visual cacophony bound together by point cloud that enables our viewing position. While it is certainly the case that the individual images that make up the synth each largely embody the conventions of three-point perspective—something inscribed in the lenses that most cameras come equipped with, and something basic to our cultural notions of visual legibility—I wish to call attention to the fabric of radical disjunctures in viewing position and the fundamentally unstable nature of the composite. In Photosynth, many points of view are called upon and inscribed into an algorithmically dynamic mix. Like Wikipedia entries, the whole only works because of individual contributions—acts of authorship, bits of information, points of view—that are algorithmically reworked into an ever-changing mix. Rather than relying on the expertise and reputation of one known individual, the reader must take a more active role in making sense of the ensuing composite of anonymous voices, in assessing it, in moving across its links to pursue additional information. The work of individuals—whether in the Wikipedia entry or in the photographic contribution to a particular synth—is theoretically traceable, but the larger mix supersedes and largely effaces those traces, enabling something like a collective point of view in which the user's agency and actions are paramount.

The organisational logic of the synth depends neither on any one person's perspective (*à la* Canaletto) nor on multiple people's perspectives (even though these are the imagistic building blocks of synth), nor even on the navigational prowess of the end user. Instead they depend on an algorithmic intervention. The selection of which images are in or out, of how they are spatially arranged, of which points offer access to which image, all reflect the interplay of algorithmically determined points of uniqueness, on the one hand, 'skinned' with photographic information, and on the other, 'activated' by the navigational work of the user. This intervention stands in sharp contrast to Canaletto's painting, which embodies the subject–object relationship emblematic of the modern era, the relationship charted by Descartes and manifest in Heidegger's *Welt-bild*. The viewing subject inscribes her position, her subjectivity, in the calculus through which the object is apprehended and represented. Three-point perspective is the technique by which this relationship is visually expressed; it serves as the language of this relationship and the guarantor of its stability. Photosynth, by contrast, is enabled through an algorithmic intervention in the subject and object relationship. A process of neither the subject's nor the object's making determines point of access, sights seen, connections made, experience gained. This is done, in a manner of speaking, in collaboration with both

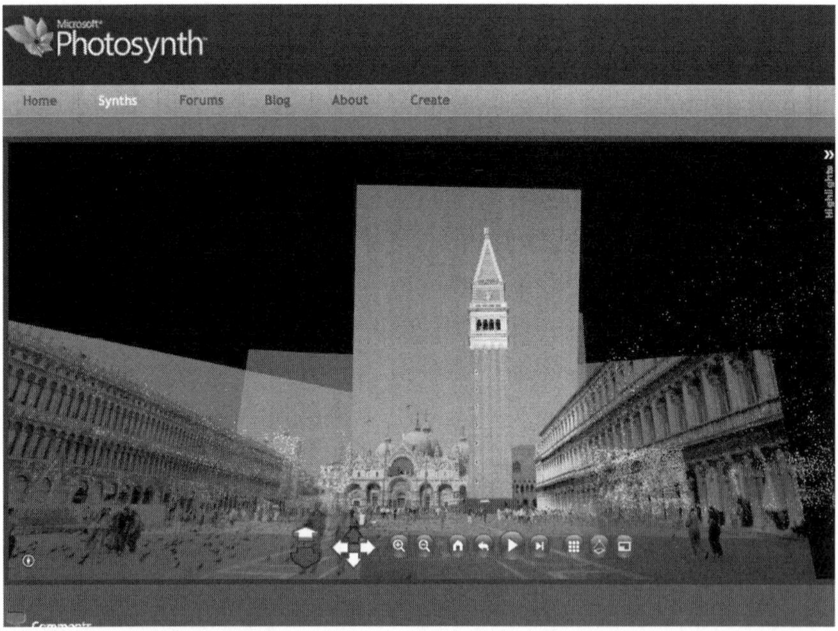

Figure 2.2 Photosynth image of Piazza San Marco, Venice. Composition by author.

the subject and object. But the effect undermines the unity of the viewing position hardwired into Canaletto's painting, in which we see the Piazza from Canaletto's perspective. In Photosynth's case, while we are aware of seeing through many complimentary sets of eyes, the enabler of those viewpoints and of the larger composite remains unseen and out of reach. It is a program layer that changes, is redefined, offers different affordances—in addition to being the single conduit through which we can access the image—while being completely outside the control of the user.

Authorship, in this context, is both problematic and pluriform. Although mostly effaced, it is the author of the individual photos; largely enacted, it is the author of the experience—that is, the navigating user; fundamentally enabling, it is the author of the algorithm; and in terms of what we actually see and select from, it is the algorithm as author. Descartes' triumphant subject and the *Ich* implied in Heidegger's *Welt-bild* are not eradicated, for their traces remain abundant. Rather, they are fundamentally repositioned by the algorithmic regimes that now stand between subject and object.

IMAGES IN MOTION: AUGMENTED REALITY

Algorithms take many forms, and no digital camera or computer-based viewing system would be complete without them. But I would like to turn to a particular application that contrasts in important ways with Photosynth: location-based augmented reality (AR). AR systems overlay existing physical reality with an additional (augmented) layer, making visible information that can only be seen through a lens or on a screen (viewing technologies are quickly developing, and although camera-equipped cell phones and handhelds currently predominate, head-mounted displays—glasses—are already on the market and working prototypes of retinal displays are now in the laboratories). The informational overlay is tagged to particular places in the world, and at the moment, three different systems may be used to link the real and the virtual. First, *fiduciary markers* are the most basic, consisting of graphically coded tags that physically attached to the object for which an overlay is sought. The device's recognition of the code triggers the appearance of the data layer in its screen, like the other systems, treating the information or image as spatially substantial (one may, in effect, walk around a virtual object as if physically exists). Second, currently the dominant application, *digital compass tracking* systems essentially offer a kind of triangulation process using geo-positioned data and GPS (Global Positioning System), a compass and an accelerometer. Wikitude, for example, calculates users' positions by using these elements, linking them together with the Wikitude data set to provide geographic information (e.g., longitude and latitude), history and information regarding points of interest. Finally, *natural feature tracking* systems represent a fast emerging image technology that assigns data to location by making visual correlations between physical places (i.e., 'recognising' them) and the information to be appended. An image-recognition system, it requires the user to position sights within a viewfinder, which the algorithms then process to find any correlations with the stored database. The system's search for unique identity points is conceptually related to Photosynth's, except that in this case, the user is in the physical world attempting to correlate real and virtual data in order to trigger a virtual graphic overlay. For the moment, natural feature tracking's intensive processing demands have limited its use to powerful handhelds and relatively iconic locations, but among industry insiders, it seems to be the application of choice.

Augmented reality applications such as Layar (http://www.layar.com/) and Wikitude (http://www.wikitude.org/), even though they use digital compass and not natural feature tracking, illustrate the uses of AR. They can overlay dynamically responsive information in the screen of the handheld; point out the locations of any particular sort of business, displaying them as overlays in the device's screen, linking those locations to web-based information repositories and providing travel instructions; give visual access to a world of 3D user-generated fantasy characters that co-inhabit real space; and offer data overlays regarding underground infrastructure such as water, sewage and electrical conduits. The AR sector has experienced exponential investment growth over the past year or two, since most smart phones now have the requisite navigation devices and AR applications turn on simple (and free) software downloads. This growth has also been driven by the advertising industry, which sees AR applications as 'the next big thing'. However, more creative and interventionist applications have also been on the rise, for example, giving users the ability to overlay commercial billboards with art projects or critical commentary, or to serve as tools for game play. These latter, more dynamic and even spontaneous applications point to the reason that natural feature tracking is preferred over other systems. If an art project seeks to overlay something relatively ephemeral such as Burger King's billboard ads and signs with a different graphic and logo (one of Julian Oliver's *artvertising* endeavours), the AR system must be able simply to recognise the graphic form rather than relying upon previously annotated GPS coordinates (something unlikely in this scenario).

My interest in natural feature tracking AR systems turns both on their temporal characteristics (their potential to embrace the ephemeral) and on their spatial requirements (the device-equipped human must be physically positioned in a particular way in order to get particular information). The requirements for physical positioning are quite specific—point the camera at the wrong location, or use an unusual angle, or an inappropriate time of day, and you will get other than expected results. In this, natural feature tracking AR systems share something with Photosynth's image requirements, where the assignment and alignment of points of uniqueness can only work if certain imaging rules are followed, assuming that the AR user has specific goals in mind and is not 'just exploring'. Unlike Photosynth, however, even if the user is 'just exploring', the resulting information in AR systems is quite specific, with meaning assigned to particular locations and information provided that can serve as incentives to act (in the case of retail businesses, coupons might appear, contact information, store hours and location instructions).

The act of mediated looking in AR systems, that is, looking through the camera of the handheld computing device, is also always an algorithmically enabled navigational act. Through the lens of the AR-activated device, there is no such thing as an innocent gaze: The act of gazing and the views consequently seen are transformed into a process of signification as images are laden with particular meanings. Paul Virilio has described a related, if far more extreme, phenomenon in his characterisation of sight-activated missile guidance systems, where seeing is synonymous with targeting (Virilio 1989). The innocence of uninformed exploration—something hypothetically possible with Photosynth—is transformed into an encounter with the always already marked and significant. The user sets the parameters of signification, be they social (are any of my friends around?), cultural (tourist information once found in travel guides is now location-bound) or commercial (where's the nearest Starbucks?).

The differences between the two systems are striking. In the case of Photosynth, we move within a virtual image space generated by the algorithmic assemblage of many points of view, many authors and even many times. And we do so presumably from a position of relative bodily stasis, sitting behind our computers. In the case of the AR systems under consideration, we move in the physical world, our movements and positioning tracked and algorithmically processed. The handheld device, our interface to the world, is both the agency of that tracking and the means of the augmented world's appearance. Point of view is embodied, constant and synonymous with the viewing subject. Rather than wandering through a virtual image space made up of many different points of view, we bridge the physical world with a virtual image space that embodies one point of view: the algorithmically modulated viewing subject. Both Photosynth and the AR applications at hand require that we take the dynamic character of the image as given. Whether lacking a dominant point of view (Photosynth) or embodied in a particular point of view (AR), movement is an expected experiential dimension: movement within the image (Photosynth) or the image in motion (AR). Movement in Photosynth is limited to the virtual world, but opens up the possibility of 'seeing through' many different points of view, through other authors' eyes. By contrast, movement in the AR domain is sited within the physical world, embedded in a multisensory environmental mix of sounds, smells and presences, but point of view is ultimately limited to that of the viewing subject.

The construction of meaning, too, functions quite differently between the two systems. Photosynths enter the world framed by meaning in the sense that they bear descriptions, are reinforced and built by image tagging, and depending on their topic, bring with them the cultural baggage of the ensemble—whether Piazza San Marco or Obama's inauguration. That framing gives way to an experiential domain in which countless other vistas and associations may be generated, since the system gives access to up to 600 or so points of view, plus the interpretive frames and interests that the user brings to bear in interacting with them. The 'steering' mechanism is simply the interaction among the user, the coded source material and the enabling algorithm in the intervening data organisation and navigation system. In the case of location-based AR applications, meanings are as precise as the viewing position. Places are dynamically tagged with particular data almost simultaneously with the act of viewing, and these tags can in turn be linked to deeper data repositories (websites) and even other links (telephone connections), reaffirming the assignment of meaning to place. AR systems effectively overlay the viewer's access to the physical world with specific (and selectable) grids of signification.

Yet for all of their differences, applications like Photosynth and augmented reality share a fundamental realignment of subject–object relations thanks to their algorithmic processing layer. As stated earlier, this does not preclude human agency, nor the stubborn fixities of the world viewed. Rather, it resituates them, defining their parameters and enabling their interactions. And although difficult to 'see' (after all, we attend to the images before us and not the underlying selection and organisation process), the algorithmic domain ultimately determines *what* we see, and even *how* we see it. There is a great temptation to rearticulate this as a form of 'old school' ideology, a distorting screen between us and the world, but this would be incorrect. For starters, it would imply the existence of an ideology-free past, somehow claiming that when subjects and objects met directly, unencumbered by the unseen hand of the algorithm, the world was a transparent place and its rule set visible to all. The long modern era of the algorismic was no less ideologically complicit. The flattening of the algorithmic and the ideological in the case of repositioned subject–object relations would also miss Heidegger's point about the *Welt-bild*. The fundamental reworking of our position as subjects vis-à-vis the world in this new algorithmically enabled era, with all of its many affordances and challenges, may well in itself suggest aspects of an emerging new order, one in which processual intermediation repositions the old certainties. We are moving from the *Welt-bild* emblematised by Canaletto's Piazza San Marco to the very different logics and subject positions of Photosynth, Google Street View, augmented reality and their ilk.

PERSPECTIVE

Let us return to Culler's notion of the marker—'any kind of information or representation that constitutes a sight as a sight, giving information about it, representing it, making it recognizable'. We might also say that it constitutes a site as a sight, transforming space into place. And it is a site of value. In the two rather divergent algorithmic applications that we've considered, we have seen different approaches to the economy of the image. In one case, linguistic tags ('Obama's inauguration') begin a sorting process that is carried on by algorithmic analysis of the image, and its dissection and reassembly as a navigable point cloud. Human tagging and algorithmic processing both mark and construct a coherent image space, with Photosynth aggregating different points of view and enabling the user to wander among them. In the other case, locations in the physical world are virtually marked and given significance. Augmented reality systems create a window into this rich overlay of data and meaning, revealing the locations of banks and pharmacies, apartments for rent and T-mobile hot spots, virtual artworks and location-based historical data. The algorithmically enabled interplay between the viewer's position in the physical world and this virtual information layer is transformative, creating sites of meaning and enabling action.

One is tempted to call upon the distinctions drawn between travel and tourism that Culler charts: one, an open and unstructured experience, with signification arising from new and unexpected ways of seeing; the other, carefully charted and well labelled, fixed in semiotic import, with the path from one sight to another well marked, and commodification often lurking nearby. But Culler is too intelligent a critic to settle for such essential definitions. He complicates these distinctions, pointing instead to shifting regimes of definitional referents, to an ages old dynamic that elevates one experience at the expense of the other, destabilising any intrinsic meaning and pointing instead to a larger dynamic. Both modes of engagement have historically struggled over the notion of the authentic. Deployment patterns of the term suggest that its meanings are highly relative, with the pot calling the kettle black when it comes to authentic encounters vs. tourism. Definitional relativism aside, times have changed. The notion of the authentic in the age of the modern, where subjects and objects encountered one another in relative peace, should probably be reconceptualised in an age where algorithmic interventions have modified the agency and place of both. The project of postmodernism might be read as the beginnings of such an endeavour. Culler reminds us that the authentic is not something unmarked or undifferentiated; rather, that authenticity is a sign relation. And in so reminding us, he offers a possible way out of this dilemma. Culler says,

> A semiotic perspective advances the study of tourism by preventing one from thinking of signs and sign relations as corruptions of what ought to be a direct experience of reality and thus of saving one from the simplistic fulminations against tourists and tourism that are symptoms of the touristic system rather than pertinent analyses. Tourism, in turn, enriches semiotics in its demonstration that salient features of the social and natural world are articulated by what Percy calls 'symbolic complexes' and its revelation of the modern quest for experience as a

> quest for an experience of signs. Its illustration of the structural incompleteness of experience, its dependency on markers, helps us understand something of the nature of semiotic structures (Culler 1990: 9)

Perhaps the algorithmic can benefit from a similar insight. Although of a different order than the clearly defined subject–object binary that characterised the modern era for the last few hundred years, the algorithmic turn remains rooted in human experiential and semiotic practices. On the one hand, that element of human continuity might override newly emerging disruptive potentials, helping us instead to locate and assess the alterations to our established modes of interaction and being in the world, and make creative use of them. At a moment of transition, it is difficult to tell whether this is simply a default mode—the momentum of the past, or if it reveals a level of adaptive insight—part of the same fabric of imagination that creates and uses these new affordances. On the other hand, perhaps we are indeed approaching a significant turn in our organisation and use of data, and these new applications can both demonstrate and enable an underlying collaborative symbolic and experiential domain outside the tradition of the transcendent subject. The implications of either stance for our understanding of the visual, its relations to the subject and the parameters of its valuation and significance, remain profound.

If the algorithmic reworking of subject–object relationships were simply limited to Photosynth and AR, the response would be simple. But the structural similarities in terms of algorithmic intermediation, a reconfiguration of subject–object relations and new dynamics for the generation of meaning and value that can be found in such diverse applications as Wikipedia, dynamic cartographic interfaces, taste recommendation systems (Amazon), Google AdSense, even automatic stock market trading, all suggest that something larger is at stake. These developments are but fissures in the still robust ramparts of the modern. The stable subject position inscribed into three-point perspective and emblematised by Heidegger as the *Welt-bild* retain their cultural taken-for-grantedness. The arithmetic precision of the algorismic remains the basis of our currencies and economic lives. But as we explore the new affordances of the algorithmic, and as our capacities to deploy them grow in tandem with the progression of Moore's Law, we might also begin to reflect more critically about the differences in emerging modes of representation. If nothing more, these differences offer compelling heuristic entry points to interrogate our assumptions regarding representation and to reflect upon the implications of the alternatives.

ACKNOWLEDGEMENTS

Originally published by *Visual Studies*, 26(1), 25–35, 2011. © International Visual Sociology Association reprinted by permission of (Taylor & Francis Ltd, http://www.tandfonline.com) on behalf of International Visual Sociology Association.

NOTES

1. 'Nothing is more boring than an unnamed landscape' (Culler 1990: 6).
2. A long and interesting history of efforts to work with algorithms predates the computer era. The work of Leibnitz and Harsdörffer, among others, greatly complicates the simple bifurcation I have heuristically asserted in this essay. For a compelling discussion, see Whitney Trettien (2009).

REFERENCES

CNN.2009.TheMoment.Retrievedfromhttp://edition.cnn.com/SPECIALS/2009/44.president/inauguration/themoment/. Last accessed 2010–06–15.

Culler, J. 1990. *Framing The Sign: Criticism and Its Institutions*. Norman: University of Oklahoma Press. Retrieved from http://www.scribd.com/doc/36984937/culler1. Last accessed 2010–06–15.

Harsdörffer, G. P. 1971. Poetischer Trichter; die Teutsche Dicht—und Reimkunst, Ohne Behuf der Lateinischen Sprache [A poetic funnel for infusing the art of German poetry and rhyme in six hours, without benefit of Latin language]. In *VI. Stunden Einzugiessen*. Hildesheim: G. Olms.

Heidegger, M. 1938/1977. The Age of the World Picture. In *The Question Concerning Technology and Other Essays*, translated and edited by W. Lovitt. New York: Harper Torchbooks, 115–154.

Ingold, T. 2000. *The Perception of the Environment*. London: Routledge.

Leibniz, G. W. 1956. On the General Characteristic. In *Philosophical Papers and Letters* (2nd ed.), edited and translated by L. E. Loemker. Boston: D. Holland.

MacCannell, D. 1976. *The Tourist*. New York: Schocken.

Photosynth. Retrieved from http://photosynth.net/about.aspx. Last accessed 2010–06–15.

Pink, S. 2009. *Doing Sensory Ethnography*. London: Sage.

Trettien, W. 2009. *Computers, Cut-Ups, & Combinatory Volvelles: An Archaeology of Text-Generating Mechanisms*. MSc thesis, Massachusetts Institute of Technology. Available from http://whitneyannetrettien.com/thesis/. Last accessed 2010–06–15.

Virilio, P. 1989. *War and Cinema: The Logistics of Perception*. London: Verso.

3 The Compact Disc and Its Culture

Notes on Melancholia

Eric W. Rothenbuhler

The year 2008 marked the twenty-fifth anniversary of the launching of the compact disc, or CD, as a consumer technology, as well as the sixtieth anniversary of the mono microgroove LP, or long-playing record, and the fiftieth anniversary of the stereo LP; it was also approaching the ten-year anniversary of the rise of the MP3. That coincidence provided an occasion for comparative thinking about those audio technologies, their places in a history of audio technologies and their contributions to our larger audio culture.[1] In fact, the past several years have seen a marked decline in the status of recorded music as a form of culture, entertainment and media; attrition in the volume of sales and thus the size and wealth of the industry; as well as, arguably, a decline in attention to sound quality and fidelity across audio media. While the CD is being displaced as a consumer medium by downloaded MP3 files, LP and turntable sales are actually increasing. This essay reviews these trends and argues that they are connected, that the CD was a paradoxical medium that shifted the value terms from culture and fidelity to cost and convenience, and was then doomed to lose a contest on those terms to the reduced fidelity medium of the MP3. Further, that shift in values and the structures of attention characteristic of the LP, CD and MP3 as media has spread to other audio media and affected the place of audio quality in the culture as a whole.

INDUSTRY STATISTICS IN HISTORICAL PERSPECTIVE

Recorded music has been centrally important in American and many other national cultures for many decades, a major medium of home entertainment and a successful industry. From the origins of the industry in the late 19th century to the turn of the 21st, recorded music has claimed its share of public attention, and other than the Great Depression of the 1930s, and a temporary downturn in the late 1970s and early 1980s, more consumer dollars with each passing decade, even adjusted for inflation (Rothenbuhler and McCourt 2004).

Since 1999, though, total recorded music sales in the US have been down nine of ten years and are now back to 1988 levels, even without adjusting for inflation. Counting units sold rather than their dollar value, the numbers are worse; album sales across formats (CDs, downloads, etc.) had a small peak in 1999, but had been more or less flat since 1994, declined nine of the ten years since 1999, and are now back to mid-1970s levels. (Sales statistics reported here were compiled and recalculated from Recording Industry Association of America reports available annually and reprinted in the Statistical Abstract of the US; e.g., RIAA 2009, 2010. 2009 data were the most recent available; preliminary reports on 2010 indicated another down year, e.g., IFPI 2011; Sisario 2011. Those reports were later confirmed, RIAA 2011, while the current news on 2011 indicates the declines may have leveled off.)

The number of CDs sold in the US hit its peak in 2000 at 942.5 million and has declined eight of the nine years since. In 2009 CD sales were down 70% from their peak in both units and dollar value, back to 1990 values of less than 300 million copies and about $4 billion. Still impressive numbers, but unlike previous patterns, this sales slide began in economic boom times and has continued across both growth and recession years. In the concert business, too, reported growth in income in recent years has masked declining ticket sales with more rapidly increasing prices. Estimates of the time consumers spend with recorded music also peaked in the late 1990s and dropped eight of the next ten years (U.S. Census Bureau).

There are two or three trends combined here: decline of the CD, the decline of music sales overall and perhaps a decline in the status of recorded music as a centre of attention and medium of culture and entertainment. Download music sales are growing quickly, and press reports of consumers switching from CDs to downloaded MP3 files are prevalent. But downloads are not growing as quickly as CD sales are declining, and preliminary reports on 2010 indicated that growth was slowing (IFPI 2011; Sisario 2011). While the industry has reported nominal increases in unit sales since 2003 (e.g., RIAA 2009, 2010), that depends on counting everything from ringtones to CDs as one equal unit. Underlying that growth in unit sales is one of the more striking changes in the industry: For the first time in decades, singles sales exceeded albums in 2007 and have each year since. Singles have grown from 1% to 2% of unit sales in 2001–2003 to 16% of the market in 2004, the first year download sales were reported, to 75% in 2009—a startlingly rapid turnaround, and a striking reduction in the total quantity of music purchased and listened to. Even at 75% of the market, though, singles represent only 20% of the dollar value of recorded music sales. Across formats downloads and physical media each constituted about 50% of unit sales in 2006, and downloads have grown to 80% since, but only 30% of the money. Physical media still constitute 70% of the declining dollar value of sales overall.

Meanwhile, the vinyl LP, predecessor of the CD, has never disappeared and sales have grown in recent years. Declining to just over a million units valued at $10 million in 1993, sales have continued at 1 to 3.5 million units per year ever since, while their dollar value has grown to $60 million. These are small numbers in the context of the industry, but steady, and they have grown each year from 2006, even through the most severe recession since the 1930s and while the rest of the industry was shrinking. From its low point LP sales have grown 350% and their dollar value almost 600%; consumers are buying more LPs and spending more per LP. (Final 2010 data available only after this was written show a one-year increase of 26% in LP sales and 44% increase in dollar value, for the best totals since 1989–90, RIAA 2011.)

The consumer electronics industry also reports increased manufacture and sales of turntables and associated equipment. Linn, one of the large British hi-fi manufacturers, announced in 2009 that they would cease producing CD players in favour of digital servers and networked players, while their turntable and phono equipment continued to sell well. The annual conventions of the Audio Engineering Society included panels on vinyl records in 2009 and 2010 for the first time in several years. Seen in historical context, this is a striking phenomenon. It is as if sales of acoustically recorded phonographs and hand-crank players were increasing in the early 1950s, people paying extra for the privilege, and engineers reconsidering the virtues of the medium. It is as if 78 rpm records and players were still being produced and winning new devotees at higher prices in the mid-1970s.

Here we have a convergence of three events unprecedented in the 120-year history of the recorded music industry. A new and ostensibly superior medium, the CD, did not eliminate its predecessor from the market place. That superior medium is now in competitive danger from a newer and sonically inferior, though more convenient, medium. Across those same years, public interest in and willingness to pay for recorded music has been declining—except for the few still using that older medium. My argument is that these facts are connected and can be explained by reference to the paradoxical nature of the CD as a medium of audio communication.

STATEMENT OF THE PROBLEM

The CD was presented and received in 1983 and the next several years as a breakthrough technology. It was fundamentally different than any of the media for home playback of music before it—different in engineering, looks, use and sound. It was digital, and the very word was new and exciting. It performed better on some of the standard audio measurements like signal-to-noise ratio and channel separation. It had a different quality of sound at the high and low ends of audio bandwidth, seemed to be flatter in frequency response and, in some senses of the word, more dynamic. It sounded different.

Yet if we look back on it now, especially in comparison with 1948 when the microgroove LP was new and 1958 when the stereo LP was new, the turning point we see in audio history is not necessarily an upward inflection. One can argue that up to and including the CD, new developments in audio technology were mostly designed, implemented and offered to consumers as improvements in sound quality, as opportunities for greater fidelity, for the superior reproduction of music in the home. But that after the CD most developments in audio technology, and all of the most successful ones, have focused on increased convenience or portability, reduced size or cost, and increased flexibility and ease of use, irrespective of audio quality. Advertising slogans capture this well. The CD was presented as a breakthrough technology; 'perfect sound forever' some ads trumpeted. MP3s, portable music players and so many of the rest of the new audio playback technologies since then have been advertised as near-CD quality. Once we were encouraged to believe in perfection forever, and now we are asked to be excited about something almost as good as what we already have.

Higher fidelity vs. greater convenience and portability is an old and enduring trade-off in audio media. In the 1920s portable phonograph players were advertised allowing music to be heard anywhere and sold alongside the top-of-the-line cabinet style players featured in high-fidelity promotions (e.g., Millard 1995; Suisman 2009). The revolution in the radio industry and the explosion of rock and roll in the 1950s depended on the popularity of portable radios, car radios, and shirt-pocket-sized transistor radios (e.g., Rothenbuhler and McCourt 2002; McCourt and Rothenbuhler 2004). Cassette tapes, more portable but with less fidelity than LPs, were extremely popular and took a substantial portion of recorded music sales in the 1980s and 1990s. All of these existed alongside higher fidelity systems and were understood as portable substitutes for better sounding machinery. In the case of the cassette, it was originally developed for business dictation; only after consumers began using it for music did the industry follow. The recorders and players that could wring more fidelity from that limited medium were packed with engineering, expensive, and advertised as such. The Nakamichi 1000 was named for its price and advertised as engineered for fidelity.

In the last decade, though, the success of reduced fidelity media has been at the cultural expense of attention to quality, aesthetics and the ideals of accuracy and truth. As computers, the Internet, MP3 files, portable MP3 players and mobile phones have become dominant audio media, they have cultured their own audience, as any medium does. This has had a withering effect on the quality of our audio culture, not only through the trading of sonically better for poorer equipment, but also through the training of inattentive listening habits and the inhibition of an evaluative vocabulary of quality and fidelity. Portable equipment and portable, often distracted listening have replaced higher fidelity equipment and more attentive, higher fidelity experiences not only in the flow of everyday life for many people, but also in the images and values expressed in the discourses of advertising and audio culture. Low price, small size, portability and convenience have largely displaced quality, fidelity, beauty or truth as primary terms of evaluation in the world of consumer audio playback media and devices. No company has made a name for itself with a tour-de-force of engineering that produces surprisingly higher fidelity from an MP3 player as Nakamichi did with its cassette decks in the 1970s. Indeed, as developed in the following argument, it would not be possible, so no one tries, and the public at large loses interest.

This cultural trend was unintentionally begun by the CD, which was sold as higher quality, higher fidelity, but which paradoxically worked against the promotion of fidelity in culture and experience. The stereo microgroove LP can be understood as a much-refined, much higher fidelity version of the same basic technologies pioneered by Edison and Berliner in the 19th century, so it fits in a story of 100 or more years of progress. The CD was a fundamentally different technology based on digital code and computation, and claimed higher fidelity because of its differences. Because it was based in computation, though, the fidelity of the CD was set from the start, and while there have been improvements in the implementation of the technology, the fundamental upper limit on its possible fidelity could not be changed (Rothenbuhler and Peters 1997). And it was sold to consumers as already perfected, so what else could we need? Lower prices, greater convenience and portability are attractive options. The CD introduced this logic, but on those criteria, the MP3 outperformed it. With that shift from a discourse of fidelity to one of convenience, as the industry statistics reported above and the analysis that follows indicate, came the end of a story of progress, replaced by drift and decline.

THE ERA OF THE ANALOG DISK

From the 1890s into the 1920s, competition among different cylinder and disk formats, groove designs, waxes, shellacs, needle shapes, tone arm and horn designs, and more were worked out over a mix of quality, reliability and cost issues—and patent disputes, of course. Within levels of reliability and cost, though, it was quality that won the competition. It was the pursuit of better audio quality as well as market access that motivated legal challenges to Victor's and Edison's patent claims (e.g., Kennedy 1994). Once those claims had been legally denied, all of the record companies began using techniques such as lateral cut rather than vertical grooves that provided the best audio quality and had the greatest market presence.[2]

Competition on price was important as well, and then as now price was not independent of quality; lower priced records were produced with thinner disks of lower grade materials. They produced a poorer sound and wore more quickly, allowing some manufacturers to advertise better sounding, longer-wearing disks. Other consumer appeals were based on musical content or cultural prestige. The old Sears catalogues show premium priced records listed according to their musical content and listings of discount priced records for which the primary advertising appeal is three for a dollar.

Compared to the 78 rpm shellac-coated record, the microgroove LP released in 1948 was a remarkable leap forward, yet the basic technique of a needle in the modulated spiral groove of a rotating disk represented continuity back to the 19th century. The new vinyl plastics were more workable taking much smaller grooves that held more information; smaller, more sensitive styli were developed to trace that groove resulting in higher fidelity signals. The records were unbreakable, quieter when new and more durable in use (though there were competing claims about how shellac and vinyl handled scratches). LPs were also more reliable in manufacturing and easier to warehouse and ship, resulting in business advantages independent of any other criteria. The microgroove architecture had enough bandwidth that the record could be turned at less than half the speed of the 78 and sound at least as good.[3] Between the microgrooves, the slower speed and a moderately increased size, the LP could carry six or more times as many minutes of music as a 78, allowing a more realistic presentation of longer forms of music.

The stereo LP was an advance over the microgroove that depended on cutting and playback cartridges that could transcribe in-phase and out-of-phase information from the two groove walls to generate right and left channel signals. In a properly set up playback system, this produces the illusion of space in the musical performance, and thus a greater degree of realism to the recorded sound. Indeed, because human hearing is sensitive to subtle cues of location and distance that appear as differences in frequency response, phase and reverberation, carefully recorded stereo LPs played back on a carefully set up stereo system produce a clear sense of three-dimensional space, the equivalent of an audio trompe l'oeil. That this works in a mechanical system of little hills and dales on the sides of grooves moulded in plastic that is generally similar to that worked out in the 1870s and 1880s, and that the modulations in those grooves vibrate a little jewel on one end of a tiny teeter totter with an electro-mechanical device on the other end to convert those vibrations into alternating current in a system generally similar to that worked out in the 1920s, and yet that the resulting sound is of such greater bandwidth, dynamics, accuracy and complexity is plainly startling. Indeed, the microgroove LP and stereo LP were received as new technologies that replaced the old. The 78 was eliminated as a medium for anyone but collectors of old records within ten years of the introduction of the LP. New mono records were gone within ten years of the introduction of stereo.

While microgroove vinyl records and stereo LPs represented dramatic leaps forward in the quality of sound for the home playback of music, each step was preceded and followed by steady improvement within the then dominant technology—as had been the case since the invention of recording in the late 19th century. The best of the acoustic recording techniques, record materials, groove architecture and phonograph playback devices of the early 1920s resulted in sound quality far superior to that of earlier decades. Development was also directed at decent sounding budget materials, records and machines so that by the middle and late 1920s, the phonograph itself was a far more popular and widespread device than it had ever been before. The development of electrical recording in the mid-1920s marked a step up in bandwidth, dynamics and naturalism—and was followed by years of steady improvement. Engineers had to learn how to take best advantage of the new capabilities, and phonographs had to be reengineered for best playback of the new records. And so it continued. The microgroove LP required a stylus with a tip one-third the size of the smallest 78 styli. Having been engineered for that purpose, though, the smaller styli were then found to be more sensitive. Building amplifiers to take advantage of that revealed new layers of sound, if the speakers were in their turn able to reproduce it, providing competitive motivation for the design of ever more sensitive and accurate systems. The stereo LP required two channel amplifiers and two properly sited speakers to produce the sound for which it was designed. The chances of incorrect playback were then more than doubled, and it took many more years for basic ideas about how to use a stereo to diffuse in the population of users—though diffuse they did, resulting again in steady improvement.

Over these same decades when the technology of home audio showed steady improvement from the phonograph to the 78 to the LP to the stereo LP, so was progress similarly measured in other domains of audio industry and culture. The sound quality of the telephone, movies and radio all steadily improved. Juke boxes too used to advertise their sound quality. Indeed, the companies behind these industries and institutions of audio quality were related and drove each other on, engineering for competitive edge (Millard 1995 makes a point of this).

THE ERA OF THE CD

The CD was promoted with great fanfare when it was new in 1983 and for the next several years. It provided one new standard for all providers for a consumer-grade digital audio medium. It was the first genuinely new medium for music playback in the home since the LP in 1948. It offered record companies a chance to resell back catalogue and a chance to bump the price of a new release over the $10 barrier—new LPs had been stuck at $9.99, despite inflation, for a number of years. It was also smaller and lighter, thus cheaper to ship than records, and involved fewer quality control problems. Equipment manufacturers and retailers got the opportunity to sell new players, the addition of a whole new category of equipment. Stereo magazines also had plenty of new material to write about and plenty of new advertising to support it.

The new CDs offered a distinct improvement over LPs in terms of some of the standard measurements of audio equipment. Their signal-to-noise ratio was larger and in principle that enabled a greater dynamic range. They offered greater channel separation. They were free of the need for RIAA equalisation and its attendant errors; in principle, then, their frequency balance should be flatter. They were simpler to insert into a stereo system than a turntable, requiring only an open line-level input; since then, line-level only pre-amps have become standard, and phono pre-amps have become optional upgrades or separate devices. Their playback mechanism was hidden from view and so different from the turntable or the tape deck, which require the active participation of the listener, they almost seemed to be free of mechanical playback at all. Certainly they were free of wow, flutter, cartridge resonance, acoustic feedback and other difficulties attendant to the mechanics of turntables and tape decks.

Neither the CD nor the CD player needed to be handled in the careful ways that records and turntables did. Early demonstrations of the new technology often involved playing dirty, scratched CDs to show that the dirt and the scratches were not audible. This is what gave rise to the advertising slogan 'perfect sound forever'. Consumers did not like the fact that LPs wore and that, if they were not handled with reasonable care, the dirt and the scratches on them were audible—later they discovered that a CD scratched badly will not play at all.

The CD was new and exciting and that was important to its appeal. It was sold as the future. It was one of the first consumer products prominently labelled digital at a time well before home and office computers had become normal. The word itself was exciting and came to be applied to all sorts of things that were not digital at all: Amplifiers and speakers were advertised as 'digital ready' and even radio stations that owned CD players were, briefly, calling themselves digital.

The CD was completely different than the thing it replaced and futuristic looking beside it. It was small and shiny; the coded side had an inscrutable iridescent shimmer. It came in something they called a jewel case for no reason anyone ever explained. And it sounded different too: brighter, cleaner, with more channel separation, maybe more dynamic. Because a good sounding turntable is expensive to buy and requires careful set up and maintenance, a moderately priced CD player brought home and casually set down on a shelf had a good chance of sounding better—or at least brighter, cleaner, with more channel separation, and maybe more dynamic.

The CD, though, did not eliminate the earlier medium the way LPs eliminated the 78. Though the major record companies converted to CD and left consumers with little choice, aficionados continued to seek out the specialty stores, magazines and community to support their interests. Twenty-five years later many recording professionals and a strong majority of audiophiles still find that a well-recorded LP played back on a quality turntable properly set up sounds naturally musical in a way that the best

CD players seldom do. Turntables, tonearms, phono cartridges and phono preamplifiers have continued to be built and sold; new designs have been regularly brought to market, earning praise for extracting yet more musical information from old records.

THE ERA OF THE MP3

The next major landmark in our story is the MP3 file. MP3 is a system of lossy data compression to transform broad bandwidth digitally coded material into a smaller data file that meets selected targets of reduced bit numbers. The algorithm for data reduction is defined by research in audio masking, the phenomenon that parts of two simultaneous sounds will be obscured to a listener. In theory the MP3 file is a more data-efficient rendition for a near-equivalent sounding audio reproduction—where near-equivalent sounding is defined by a statistical model of a listener in a statistical model of a listening situation.

MP3 was never intended as anything other than an efficient system for transmission of digitally recorded audio and video. Since the compression is lossy, that is, it destroys data, MP3 transformation is permanent and could never be considered adequate when fidelity or quality are primary goals. Though MP3 recorders are popular for office and legal dictation, as interviewing devices, and as inexpensive and convenient field recorders for cash-strapped academics, MP3 is never used as a medium for professional recording of music.

In the right listening situation it is readily apparent that the MP3 coding of a CD is audibly inferior, as is acknowledged even by its proponents (e.g., Sterne 2006). It is rare, though, that consumers find themselves in such a situation of quiet concentration with two differently encoded copies of the same recording played back on a system that can make the differences audible. And the larger cultural discourse tells them not to worry, that they would not hear a difference they cared about anyway. Just as the MP3 coding 'pre-listens', discarding information that its model predicts would be masked, the discourse 'pre-judges' that we would not care.

The MP3 may be reduced fidelity, but it is a huge success with consumers. It is cheap or free, easy to download, takes no physical space outside the computer or player, is small on the hard drive, can't get scratched or dirty, is easy to share, and apparently it sounds good enough. In the same years that MP3 was gaining popularity, higher bitrate, broader-bandwidth disc media like DVD-A and SACD failed to catch on in the market. High bitrate downloads available now are not widely known by consumers at large. Even the readily available and perfectly convenient non-lossy compression systems available on most everyone's computers are seldom used.

Similar patterns hold for other media. Dynamic compression, for example, reduces the range of volume differences in naturally occurring sound.

It is routinely used in small amounts while recording to improve signal to noise ratios and more extensively in some styles of music as an aesthetic effect. Television and radio advertisements, though, are produced with radically compressed audio so they can be made louder. Radio stations compress the dynamic range out of their audio to increase fringe reception. TV and movie sound is compressed and manipulated too, louder and more attention getting than ever, but less realistic and arguably less beautiful. All of these are examples of audio production techniques for business strategy rather than expressive value, for impact rather than meaning (Rothenbuhler 1996).

Cell phones are replacing landlines when, by comparison, as audio devices they are essentially broken. Their performance on sound quality measures would not be tolerated anywhere else, except perhaps on Skype, because it is free, and that's more popular today than quality. At least there you can see who is talking so it is less important to recognise the voices of your friends and family. Ringtones and ring-back tunes provide a new outlet for recorded music, but they render it almost unrecognisable, and the abbreviation of musical form is incoherent and meaningless. Using a few seconds of musical sound as a substitute for 'brrring!' may be the ultimate expression of the reduction of music to social function that Adorno (1941, 1945) predicted.

THE LP, CD, AND MP3 IN USE

In use the LP requires careful handling and a properly set up and maintained player. Different copies of the same LP and different players have different qualities of sound, drawing attention to that element of the experience. Better equipment and more care in use result in better sound, thus rewarding attention to detail, encouraging fastidiousness and enabling the narrative of progress. It was designed for longer pieces of music, and the LP and its player are large enough objects that the listener tends to comport him or herself in relation to them, not vice versa. The player has to be put in its own place somewhere and the LP cover has to be set somewhere while the record plays, often propped visibly against the wall or equipment stand. It cannot be played in a computer or car, nor in a shirt pocket, cutting down on the opportunities for distracted listening. The LP comes with photos, graphic design and liner notes in a large format; the user has something to look at and something to display. Conventionally the liner notes provide additional information about the musical performances and listeners are invited to think that important. Learning more about the music is eased because the information is pushed to the consumer.

The MP3 file has no physical presence at all, cannot be handled, looked at, or comported in regard to. It sounds worse on better equipment and its defaults are only apparent when paying attention; in this way it punishes attention to detail. It can only be played in a computer or other microprocessor based device, which were not principally designed for audio and often sound like it. It is most frequently used on portable players; low fidelity and distracted listening are its norms. It comes with little or no information and players commonly limit the character count of titles, truncating them to fit the screen rather than the information. Listeners are invited to think no more information is needed and anyone wanting to learn the name of the trumpet player or which Goldberg Variation they had just heard will need to do their own searching elsewhere. In this way, while the MP3 offers many conveniences, as Bolin (2009: 348) pointed out about digital commodities in general, it requires both special equipment and extra consumer effort to realise the value that was given in the analog form.

The CD lies between these two extremes as a small object that one can collect and display, but that requires less care, is less encouraging to fastidiousness, and usually offers less information than the LP, but not none. While most consumers only had one turntable and would usually go to the room where it was to hear a record play, the CD player began to proliferate, appearing in cars, portable players, clock radios, DVD-players and computers. The CD moved to be heard where the consumer was. In the chronological account it appears likely that the CD, and the cassette as well, started the trend that the MP3 culminates. It was not just the size and portability of the CD that mattered though. Different copies of the same CD sound the same; different players do not sound as much different as different turntables, tonearms, and phono cartridges do. The differences between the average and the best players are smaller than with turntables; progress in sound quality has been less marked. The CD player introduced random shuffling as a way to listen to music, perfected in the MP3 player where any song is treated as equivalent to any other. The CD made indifference a viable consumer attitude.

THE CD AS A MEDIUM FOR MUSIC

These characteristics of the CD in use are catalysts for a problem inherent in its logic as a medium for recorded music. The CD was the advent of digital playback in the home and compromises of its particular implementation of digital coding turned out to be key.

Digital recording and playback are systems for the measurement of sound; the coding, storage and decoding of those data; and the synthesis of an alternating current signal analogous to the alternating current signal that was originally measured. Like all data systems, digital recording and playback require a set of a priori assumptions about the range, precision and volume of the data and an agreement about coding and notation procedures. Should we measure and store data representing precision of 0.1, 0.09, 0.099, or 0.09999999? Each choice in the sequence takes more space in the storage medium and could only be used by a decoding algorithm that anticipated that level of precision. As a medium for data, the CD is designed to hold a defined amount of data of a defined precision encoded by a recognised convention so that it can be decoded by standard procedures. All of this was worked out in inter-corporate negotiations and has since been referred to as the Red Book standard. Crucially important was a compromise that selected a bitrate lower than used in professional recorders in order to squeeze more minutes of music onto a disk size selected for portable and car players.

The data system standards of the CD are straightforward enough. But the CD is not, in social use, a database or an information system. It is a medium for the home playback of music. So the limits on the data became limits on the fidelity and quality of musical sound. Of course, no one's stereo is perfect; all playback media and devices work within practical limits. But there is a fundamental difference between the logic of the limit on digital fidelity and the logic of the limit on analog fidelity (Rothenbuhler and Peters 1997).

Analog recording and playback are problems in mechanical craft; digital recording and playback are problems in calculation and data storage. So the limits that analog devices impose on fidelity are based in their mechanism or their use, neither of which need be assumed to be anything less than perfect, and both of which are always improvable. Of course there are elements of compromise in every real device—and in the world of consumer products designed for mass manufacturing to distinct price points, the compromises are explicit in the engineering. Also of course there was craft in the original design of the CD standard, in the development and implementation of different coders and decoders, in digital recording, CD mixdown, and so on. The point here is that the analog medium does not *require* an a priori decision about the upper limit of achievable fidelity, only a decision about how much to spend pursuing it. Just as any carpenter guides his or her work with concepts of perfection such as straight, square and plumb, while ignoring small deviations unimportant to the scale of the project at hand, analog media can be designed with the idea of approaching perfection, even as they fall short, and yet remain open to the possibility of improvement. A digital medium, though, as a data system rather than a mechanical device requires that its finite bounds be specified and designed in, after which they can never be exceeded, no matter how good the craft becomes.

So, when inter-corporate negotiations settled on the sampling rate, bit rate, encoding conventions and size of the CD format, they set a hard upper limit on bandwidth and data precision—thus also on the goal of fidelity and the dream of obtainable truth. Improved CD players, even a theoretically perfect one, can sound no better than the compromises made in the design of CD encoding. By contrast, analog recording and playback are based in mechanical processes rather than measurement and coding designs, so the practical compromises of a given medium do not establish hard limits on fidelity in principle. This leaves something to be discovered and the dream of perfection still alive. New and improved turntables, for example, have increased the appreciation of the information available in the microgrooves of old stereo LPs, demonstrating that the real limits of fidelity were not yet known and providing motivation to continue trying. By contrast, while CD players improved rapidly in the first ten to fifteen years, progress since has been more like an asymptotic approach to a limit.

CULTURAL IMPLICATIONS

When evaluating the cultural implications of a technology, Innis (1951) recommended we look for the paradox of strength and progress, for the inbuilt logic of the success of a technological-cultural complex that foretold its own decline. The implicit logic of the CD is that we already know what we need to know, we already know how to achieve our goals, we already know that things cannot get any better than this. There are, in the world of the CD, no mysteries. Regardless of the practical limits on fidelity and the fraught status of truth, with an analog medium we can imagine the true, dream of perfection and strive for the better. With a digital medium all of those questions are already answered: This good and no better. So while the experience of the audiophile with a record player has a structure parallel to that of the experience of religious revelation—greater care and attention produces nearer approach to perfection—the experience of the digital audio consumer is just that, the experience of a consumer, conducted by the logic of consumption. Further, in that digital world where we already know the answers to our questions, the consumer who optimises cost and convenience is being rational, and the audiophile who invests more for better sound from the same recordings is pursuing chimeras.

The dominance of the CD contributed, then, to a general disparagement of the idea of fidelity and thus the dreams of truth and perfection. With the elimination of such ideals from the discourse of home playback of music, nothing was left to guide our choices but efficiency, convenience and price. In that context, judged by those standards, the MP3 file is indeed a great thing. But if it contributes to a world in which no one can hear or wants to hear that it is a poorer reproduction of the sound of music, a world in which enthusiasm for its low price, efficiency and convenience displace any attention to its poorer fidelity, then is our world at risk of becoming more efficiently less beautiful, pleasant or true? And if it is not resulting in a less attractive, less satisfying audio experience, then why is the time and treasure the public devotes to recorded music declining year after year?

With the rise to dominance of the CD came a growth of indifference in audio culture: indifference to the time and place of listening, fidelity of sound, access to more information about the music, the handling of the recording itself, degrees of attention paid, even which song played next.

That indifference led the vast majority of consumers to cheaper, easier choices despite the lower quality sound. And now the same market logic indicates they are getting less out of that experience and willing to put less into it. The paradox of the CD is that its success led to its decline; the tragedy is that it may also have lead to the decline of recorded music as a medium of culture, communication, and entertainment. The result may be a less beautiful, less sensorily rich, less thoughtfully engaged world.

NOTES

1. This essay is substantially revised from two earlier versions under different titles presented at Cultural Technologies / Cultures of Technology Symposium, School of Culture and Communication, Södertörn University and the Nobel Museum, Stockholm, Sweden, October 19–20, 2009; and at the Annual Convention of the International Communication Association, Montreal, Canada, May 23–26, 2008. My thanks to Nancy Baym, Göran Bolin, Patrick Burkart, John Ellis, Staffan Ericson, Tom McCourt, Jonathan Sterne and several other participants in the vigorous discussions on both occasions.
2. The historical discussion here is based on standard sources, with my own interpretations. See Chanan 1995; Day 2000; Dearborn 1987; Eisenberg 1987; Gitelman 2000, 2008; Katz 2004; Kenney 1999; Millard 1995; Pohlmann 2005; Sterne 2003; Strawn 1985; Welch and Burt 1994; White 1991. The essay is also informed by thirty years of reading in the audio hobbyist literature and repeats much that counts as common knowledge there.
3. Bandwidth is in part a product of speed in devices like record players and tape machines where an analog signal storage medium is moved past a transducer.

REFERENCES

Adorno, T. W. 1941. The Radio Symphony. In *Radio Research 1941*, edited by P. F. Lazarsfeld and F. N. Stanton. New York: Duell, Sloan and Pearce, 110–139.

Adorno, T. W. 1945. A Social Critique of Radio Music. *Kenyon Review*, 7(2), 208–217.

Bolin, G. 2009. Symbolic Production and Value in Media Industries. *Journal of Cultural Economy*, 2(3), 345–361.

Chanan, M. 1995. *Repeated Takes: A Short History of Recording and Its Effects on Music*. London: Verso.

Day, T. 2000. *A Century of Recorded Music: Listening to Musical History*. New Haven, CT: Yale University Press.

Dearborn, L. 1987. *Good Sound*. New York: William Morrow.

Eisenberg, E. 1987. *The Recording Angel*. New York: Penguin Books.

Gitelman, L. 2000. *Scripts, Grooves, and Writing Machines: Representing Technology in the Edison Era*. Palo Alto, CA: Stanford University Press.

Gitelman, L. 2008. *Always Already New: Media, History, and the Data of Culture*. Cambridge, MA: MIT Press.

IFPI. 2011. *IFPI, Digital Music Report 2011*. International Federation of the Phonographic Industry. Accessed at: http://www.ifpi.org/content/library/dmr2011.pdf

Innis, H. A. 1951. *The Bias of Communication*. Toronto: University of Toronto Press.

Katz, M. 2004. *Capturing Sound: How Technology Changed Music.* Berkeley: University of California Press.
Kennedy, R. 1994. *Jelly Roll, Bix, and Hoagy: Gennett Studios and the Birth of Recorded Jazz.* Bloomington: Indiana University Press.
Kenney, W. H. 1999. *Recorded Music in American Life: The Phonograph and Popular Memory, 1890–1945.* New York: Oxford University Press.
McCourt, T. and E. W. Rothenbuhler. 2004. Burnishing the Brand: Todd Storz and the Total Station Sound. *The Radio Journal*, 2, 3–14.
Millard, A. 1995. *America on Record: A History of Recorded Sound.* Cambridge: Cambridge University Press.
Pohlmann, K. C. 2005. *Principles in Digital Audio* (5th ed.). New York: McGraw Hill.
RIAA. 2009. *2008 Year-End Shipment Statistics.* Washington, DC: RIAA.
RIAA. 2010. *2009 Year-End Shipment Statistics.* Washington, DC: RIAA.
RIAA. 2011. *2010 Year-End Shipment Statistics.* Washington, DC: RIAA.
Rothenbuhler, E. W. 1996. Commercial Radio as Communication. *Journal of Communication,* 46 (Winter), 125–143.
Rothenbuhler, E. W. and J. D. Peters. 1997. Defining Phonography: An Experiment in Theory. *The Musical Quarterly*, 81, 242–264.
Rothenbuhler, E. W. and T. McCourt. 2002. Radio Redefines Itself, 1947–1962. In *The Radio Reader: Essays in the Cultural History of Radio*, edited by M. Hilmes and J. Loviglio. New York: Routledge, 367–387.
Rothenbuhler, E. W. and T. McCourt. 2004. The Economics of the Recording Industry. In *Media Economics: Theory and Practice* (3rd. ed.), edited by A. Alexander, J. Owers, R. Carveth, C. A. Hollifield and A. N. Greco, Mahwah, NJ: Lawrence Erlbaum Associates, 221–248.
Sisario, B. 2011. Another New Low Atop *Billboard* Chart. *New York Times*, January 19, 2011.
Sterne, J. 2003. *The Audible Past: Cultural Origins of Sound Reproduction.* Durham, NC: Duke University Press.
Sterne, J. 2006. The MP3 as Cultural Artifact. *New Media & Society*, 8(5), 825–842.
Strawn, J. Ed. 1985. *Digital Audio Engineering: An Anthology.* Los Altos, CA: William Kaufmann.
Suisman, D. 2009. *Selling Sounds: The Commercial Revolution in American Music.* Cambridge, MA: Harvard University Press.
U.S. Census Bureau. (various years). *Statistical Abstract of the United States.*
Welch, W. L. and L. B. S. Burt. 1994. *From Tinfoil to Stereo: The Acoustic Years of the Recording Industry, 1877–1929.* Gainesville: University Press of Florida.
White, G. D. 1991. *The Audio Dictionary* (2nd ed.). Seattle: University of Washington Press.

4 Peer-to-Peer File-Sharing Systems
Files, Objects, Distribution

Mats Björkin

> In art, the spectator, the reader, the listener is a problematic subject. Any creative spirit is completely indifferent to him. However, in an invention like the cinema, the spectator has an important role, because cinema is *a priori* something democratic, popular, created expressly for a broad public, and is so not only from a financial standpoint but also in its core essence. (Langer 1913/2008: 95)

František Langer's short ontological remarks on film from 1913 could have been made almost anytime during the history of cinema. At the same time, such remarks have been controversial throughout its history. Later in the same article, Langer likens the film medium to an event and its viewers to witnesses:

> Film is therefore an *event*, and the spectators are pedestrians in the street who have stopped to look, just as they would stop on the street to look at some interesting incident, be it a fallen horse, or ice-cutting and loading down by the Vltava river. (Langer 1913/2008: 96)

Today, the same viewers are often seen as consumers, and when they do not pay, as simple shoplifters. Film theory has spent much interest in describing the act of watching, or what Langer described as witnessing. Film economics has described and explained the economic transactions behind film production and film consumption. Less research has focused on what film viewers have actually been paying for.

Langer was, as we all know, typical of his time. The entire field of early cinema studies during the 1980s and 1990s was fed and breathed by such texts (cf. Bruno 1993; Charney and Schwartz 1995). This research was then adopted by new media studies, as a tool for the analysis of the fragmented aesthetic objects of the digital arts and the reception patterns discernable as a result of Internet (cf. Manovich 2001). What I wish to do in this text is discuss some theoretical consequences of treating file sharing as shoplifting (rather than window shopping). In addition to this, I wish to raise some questions concerning how file sharing may change our view of film spectatorship.

I use the history of film and television consumption as a tool for analysing file-sharing systems as a new form of distribution for film and television. I focus on some distinguishing features like peer/user labour—how viewers gain and construct information about content and technical characteristics of the data shared in the systems. I conclude my analysis by way of a case study of illegal file sharing of hardcore pornographic films in order to emphasize what I argue to be the most important characteristics of peer-to-peer file-sharing systems.

There are two basic rationales behind file sharing: 1) gaining access to commercially available media material without paying for it, and 2) gaining access to media material that is not commercially available and therefore not possible to pay for. The first rationale is based on an agreement that certain events are free while others are not. In the early years of cinema, the film screening itself was sometimes what you paid for, while on other occasions (and at other places) you got the film screening 'for free' when you paid for another event (e.g., a Vaudeville show). The establishment of cinema theatres gave way to, and depended upon, an agreement between moviegoer and cinema owner regarding costs for entry into the auditorium. Seen in retrospect, it was never really established whether moviegoers paid to see the film or for the physical entry into this specific space. For the sake of the argument here, I will use as a premise that moviegoers have always paid for events. They were not simply 'pedestrians in the street who have stopped to look', as Langer argued. They were included in a particular social situation in which the looking naturally may have been considered free but the space provided for the shared watching was worth paying for.

With the entry of television, the act of looking was still not connected to an economic transaction. In the case of public service television, viewers paid licence fees for the access to the transmission technology. In commercial television the economic transaction was even more indirect, in the sense that viewers paid for the access to programming through the marketing costs of goods and services. Theoretically speaking, the availability of the access to both technology and programming might have increased or decreased the costs of goods and services. If television viewing did cost anything at all, this cost was connected to a shared (or individual) event, now within the domestic space, which the viewers in one form or another had already paid for.

With the videocassette, and later the DVD, the material foundation for both the economic transaction and the event, moved from space to object—or, in other terms, from one particular disposition of objects into another. As in the case of television, domestic furniture and electronic machinery set the scene for viewing, but with the videocassette (or DVD) as object, questions concerning collecting and sharing became added. Again, it is possible to argue that we do not pay for viewing what is on these tapes, but rather for the possession of the physical object. The premise, economically speaking, is of course that these objects are scarce, they are physically discernable, and they are, as objects, unique, which amounts to each economic transaction being connected to an individual physical object.

The characteristics of digital media have been heavily discussed during the past two decades. For my argument here I will only use the technical characteristics of the digital file, which is neither scarce nor physically tangible, as a stepping stone to my main argument: a plea for the re-recognition of the relationship between the economic transaction and the event. Or of the fact that, in Langer's terms, that it is in this relationship that we can locate film's democratic potential.

MOVIEGOING AND OBJECT RELATIONS

The changes in distribution due to the Internet, both the legal and illegal distribution of audiovisual products, have inspired—or provoked—a large number of scholarly and nonscholarly texts about intellectual property rights in the shadow of the pirate, as well as about distribution. The growing body of literature on film distribution—in economics and marketing, political economy and of course film history (particularly in relation to Hollywood and the US)—all emphasises the importance of where the money is. The power of production has always been in the hands of those closest to the principal sources of the income, who during the classical Hollywood era were the theatre owners (Balio 1985: 262). Again, since these successful economic transactions have historically been connected to a particular event—*going* to the movies—the dispositions of objects are key to understanding the mechanisms involved.

Psychoanalytical film theory has taught us that the crucial moment of the film experience is the individual viewer in a dark space looking at the world represented on screen (Kaplan 1990). Regardless of how we define this act of looking—as gaze or just looking—this particular event is inseparable from the physical room, the furniture, screening technologies, the other people present even if they are invisible to each other, the doors to the room, the attendants, the box office and of course the act of buying a ticket. The 'gaze', thus, is necessarily linked to the cinema as a physical and social space, and a particular form of economic transaction. For these reasons alone, it would be possible to argue that gaze theories cannot work for television, home video or computer-based film watching.

The more general consequences for film theory belong in another study, but for my argument this opens up for looking at file sharing from a new perspective. When we watch films outside the cinema theatre, at home on videocassette or DVD, the object of our consumption no longer depends on us being part of an event. Since we are paying for an object and not an event, we most likely enter into another relationship with the film. Since gaze theories explain a certain mode of watching (within the cinema theatre) and a specific mode of cinematic storytelling (classical Hollywood narration), I find it reasonable to argue that these two phenomena condition each other. Classical Hollywood cinema gives the perfect experience for an event in a cinema theatre worth paying for. A hundred years of Hollywood success is enough proof of that.

The fragmented narration of television and digital media, elaborately theorised at least since the 1960s, obviously seems to thrive in other places than within the cinema theatre. Of course other modes of narration have also been screened and watched at cinemas, but cinemas are still dominated by the kind of narration for which the cinema as an institution was developed. For my argument here, it is enough to conclude that there seem to be certain quantitatively strong links between two distinct setups of spaces, viewing patterns, economic transactions and modes of narration.

FILE SHARING: A RE-DISPOSITION OF OBJECTS

Just a few years ago, it was possible to find articles or blogs asking questions like 'Will BitTorrent Change Television?' or 'When P2P Gets Work in Hollywood' (McKee 2007; Boudet-Dalbin 2007). One could easily get the impression that peer-to-peer (P2P) file-sharing systems, apart from being a legal problem, also (and perhaps just because of this) promised to be the future of film, television and music distribution. Four years and some Pirate Bay court cases later, peer-to-peer file sharing seems to be regarded as either a popular uprising against the commercial media industries or a lethal threat to the same industries (Sanchez 2008). Then again, more productive approaches to file sharing emerge, as criticism of the industry demonisation of file sharing but also as criticism of over-the-top techno optimists (Moss 2011).

Since most file-sharing systems, particularly BitTorrent, depend on as many contributors as possible for the efficiency of the system, there will always be a tension between individual rationality and collective welfare (Babaioff and Chuang 2007). This tension is difficult enough in illegal activities; when trying to create business models based on peer-to-peer file-sharing systems, the problems are innumerable. Searching for a business model for the peer-to-peer-based distribution of film and music is, not surprisingly, a popular quest. The fact that Spotify, at the same time as video services using similar models, like Voddler, are not as successful, is an example of the demand, on the one hand, and the fact that the main problem is legal rather than technical, on the other. Distribution has always been the key to making money on audiovisual media, not only for traditionally economic reasons (as intermediary between producer and customer) but also as providing the basis for the events. In the line of Langer's analogy, the most important thing is not only the person looking or the owner of the street where the looking takes place, but also the providers of the horse falling and the ice being cut and loaded—and, in my view, where the economic transaction takes place.

In peer-to-peer file-sharing systems, no object exists outside its relations. If no one makes a particular file collectively available, this file will only exist in individual computers but not within the system. In this sense, we are not talking about a traditional long-tail phenomenon (cf. Anderson 2006) but rather the opposite. Peer-to-peer file-sharing systems work best for blockbusters, when many contributors make the same file available at the same time. This, of course, puts emphasis on the object—the individual file—but in another way than before. If the unifying agent during the Hollywood era was based on a combination of the cinema theatre, the celluloid film object (invisible to the moviegoer) and the lights and shadows projected on the screen, the material commodity itself took over this function during the videocassette and DVD eras. The intensified attention to the object during the digital era, I will argue, has resulted in a redefinition of the relationship between object and viewer/user, actually 'referring back' to the event. The major reason for this is that the blockbuster (rather than the long-tail) feature of peer-to-peer systems depends so much on copying.

COPY AS BASIS FOR CREATIVITY AND DEVELOPMENT

As a cultural phenomenon, file sharing can also be discussed in terms of ideas, practices and values being developed and expressed through social networks. And, as in all social networks, few behaviours are as common as copying others (Ormerod and Bentley 2010).

Technically speaking, the issue of copying others is vital for peer-to-peer networks to work. As cultural phenomena, both movie going and file sharing depend more on copying others than on individual, rational choices. But, again technically, there is one major difference. Going to the movies to see a classically narrated movie is a combination of being able to share the experience with your friends and finding the films you think you will like. Copying the behaviour of people whose taste you trust minimises the risk of disappointment. But copying the behaviour of people you do not know also reduces this risk, at least if you know your own preferences in relation to the majority.

The difference between movie going and file sharing is crucial in one sense. With increased knowledge of your own preferences in relation to others', your decisions will most likely be focused on what makes the difference (for you).

So, is there a difference between movie going and file sharing in this respect? My answer is yes. I argue that there are two main reasons for this difference, both related to the relation between information and economic transactions: how to find what you are looking for, and how to know if you have found it. Both involve the investments (emotionally, technically, financially) and the delicate processes of measuring value.

As discussed by Mark Deuze, Henry Jenkins and others, there is a great deal of work being done by those who consume, interpret and rework existing media products (Deuze 2007; Jenkins 2006b). Most of the work, done today by (amateur) users by way of illegal file sharing, will be done either by media industry professionals if it is transformed into existing business structures, or, as today, by the users. Or both. On the other hand, if the use of a peer-to-peer system becomes more than a technical issue, it must be considered how to value the work of peers, or at least the use of their equipment. This is crucial because if you remove the incentives for user activities it is not only a question of lost buyers; their work and equipment may also disappear from the system. And if users and their equipment disappear from the system, the system will disappear. Traditional moviegoers provide financial prerequisites for the movie theatre. If they stop going to the theatre it may become impossible to continue running the theatre, but the theatre will remain as physical space with all its different objects.

INFORMATION, OR HOW TO FIND WHAT YOU ARE LOOKING FOR

A key issue has to be how to find the files you are looking for. Distribution determines who gets to watch films, under what circumstances and why. In a web-based system, the search function is particularly crucial. Research on search engines indicates that there is a gap between what the search engines can do and what the user can do with them. User competence has been quite low, which has led to users treating search engines uncritically, unaware of their workings (cf. Vaidhyanathan 2011). The low level of competence seems not to accord with users' self-assessments, according to a recent study (Machill et al. 2008). This competence is most likely generally increasing, but it is still unclear if the earlier dominance of one-word searches is actually changing, and at what pace. This research is based on general search engines, but since most torrent sites use similar search engines it is, at least partly, applicable here as well. This knowledge also impacts on what is being downloaded, since it is obvious that the most popular downloads are quite easy to find. However, is reasonable to believe that since torrent sites are used more by computer- *and* media-literate users than other ordinary search engines are, the competence to use a search engine at a torrent site may be above average and, more importantly, the knowledge about what they are capable of searching is generally quite extensive. The question is what kind of competence is implied by torrent sites' search engines today. Or, in other words, do we need other skills and competences here compared to, for example, when looking for a DVD or a movie theatre?

Many torrent search sites have a very limited classification system ('movies', 'TV', 'music', etc). This means that browsing for material is often conducted without the possibility of using genres or nations, or even names, as in 'traditional' media archives or catalogues of online DVD retailer catalogues. The most common browsing categories are 'most popular' and 'most recent'. For browsing, the user has to rely on titles.

Being so dependent on titles means either that torrent site search engines promote skills in knowing and remembering titles (film buff skills), or the user has to rely on complimentary databases like the Internet Movie Database (IMDB). Some torrent site search engines have developed space for including filmographic texts, which have usually been copied from the IMDB or from DVD covers. This makes all included data searchable, but it also means that files with metadata will be easier to find than those without metadata.

First of all, the film buff behaviour could clearly be seen as a variation of the tendency that early adopters dominate the use of specialised services. On the other hand, there is little research on whether torrent site users are also early adopters of technology. According to research by Papies and Clement (2008), consumers perceive movie consumption over the Internet as technically demanding, which hinders the adoption and diffusion of Internet shopping for films. This restriction partly addresses the questions of file formats and the difficulty most people experience in watching a downloaded file on the television without prior possession of certain technical skills. But it also concerns difficulties in the search itself. If this is so, file-sharing activities would be attractive to a specific, though rapidly growing, target group: those who prefer watching films and TV on the computer instead of the television (Papies and Clement 2008). As watching films and particularly television on the computer is becoming more and more popular, these two parameters may be connected, but we still do not know for certain if it concerns the same people.

At the same time there are arguments asserting that it is users with strong digital experience who are most likely to adopt file-sharing services, rather than those who exhibit a generally high level of film and television consumption. In an age of 'convergence culture', the consumers of the media industry would naturally be both compulsive consumers *and* digital natives. It is only then that file sharing would become a mass phenomenon, or a big problem—depending on perspective, although this does not yet seem to be the case.

Here, two different patterns of use influence other users' activities. Uploaders of niche products, cult films or vintage material are often interested in details: which people were involved, character names, detailed plot descriptions, and so on. This is, again, a type of film buff behaviour. If this interest results in a concern for including such information in search systems, these files will be easier to find. But, it seems that this kind of material often lacks additional information, probably because it is already known among enthusiasts. They do not need any more than the title to find what they are looking for. This reduces the number of downloads, but when the activity is limited to enthusiasts, who often have a high level of uploads, enough files are reasonably accessible most of the time. If so, this is the long tail in action. But, with very few files available the risk of failure increases: corrupted files (damaged or unplayable), corrupted data package heads (incomplete or interrupted downloads), temporary breaks (particularly at the end of the download when there are fewer available files), and so on.

The other scenario, as far as I have seen, is that uploaders of the most popular commercial films and TV shows, when possible, include the same information that is found on the retail DVD or in TV guides. This makes searches easier for nonspecialised users. The combination of popular items and useful information results in so many downloads that it compensates for regular users' tendency to remove a torrent file as soon as it is downloaded, which of course reduces the public exposure, but makes the files less accessible to other users. This is, thus, a typical 'head' rather than 'tail' phenomenon.

With metadata, peer-to-peer networks work fairly similar to commercial distribution systems, with the difference that no one makes any money. The Pirate Bay was a typical example of this kind of torrent search engine, which mimics commercial distribution and consumption. Perhaps this is one reason why these search engines have been so much in focus of both legal action and public debate. What complicates the discussion is that the same interface, for example, the Pirate Bay, combined both modes of use, or both kinds of behaviour. But arguing that there is one 'good' part (access to products which are not commercially available) and one 'bad' part (illegal dissemination of commercial products) could be misleading. Both kinds of behaviour are examples of market disequilibrium, in terms of classical economics. It is an extreme imbalance in opposing definitions of value, not only in economic terms. Before finding a market price, we have to determine whether it is at all possible to place an economic, or other, value on a product. Or, in other words, what is the product?

The traditional cinema system employed what could be described as an almost binary logic: Either the film is worth the predefined ticket price, or it is not. There is no bargaining (although there has always been some degree of price difference between theatres) and no return of ticket prices if the screening was not to the moviegoer's satisfaction—only the difference between regular cinemas (of different kinds), cinematcqucs, festivals, and so on. The most plausible reason for going to the cinema is that the audience and the cinema owners shared the evaluation of the moviegoing event. In line with this argument, does this mean that file sharers are unwilling to pay for films? The answer depends on what we mean when we talk about paying for films. Are we paying for the event or the object? I argue that to answer this question we must first decide whether it matters at all. Since movie going still occurs, most people still want to pay for the event. But this, obviously, does not entail an interest in paying for objects that are simulations of what is projected in cinemas. There seem to be other issues at hand concerning what the object of consumption really is.

HOMOGENEOUS PRODUCTS, OR HOW TO KNOW WHAT YOU ARE LOOKING FOR

A matter perhaps even more important than knowing how to find relevant viewing material is knowing what to look for. This brings us back to the question of titles. In some, very popular, torrent search sites, most searches have to be title-based. But titles, as well as possible metadata, of individual media files are given by the uploader, and are thereby dependent on choice of language, spelling, and so on. There will therefore always be uncertainties about what the file actually contains. Obviously something happens with the idea of a media product as an object or as an artistic work when it becomes increasingly difficult to predict what the media file actually includes. As mentioned earlier, skills for remembering titles or relying on other databases are crucial to the user, in order to find what she is looking for. In both cases this works against overly subjective and creative titling. On the other hand, since many torrent sites require downloaders to upload as much as they download, it encourages titles that attract other downloaders, which sometimes stimulates a certain kind of creativity.

In relation to this, two diverging tendencies can be observed. On the one hand, among the most popular titles—blockbuster movies or the most popular TV shows—'creative' titles are rare. The only differences are the release titles in different languages. The same film thus appears many times with slightly different titles, but is still easily recognizable and thereby easy to combine with previous knowledge or other databases. Again, this is the Pirate Bay model.

On the other hand, for specialised items such as rare films and TV series and episodes, which are not always accessible, creative titling can become a way of 'tricking' users to increase the download in order to facilitate a necessary level of available files. More important is that torrent search engines with limited extra information rely heavily on 'film buff skills' or access to additional databases. It is most likely unintentional, but the lack of extra information creates 'clusters' of specialists who are more likely to download a broader range of files and are more keen to make the files available for others to download.

As far as I can see, large 'commercial film distribution-like' torrent search engines like the Pirate Bay work well as pirates, but have a more limited function as a long-tail opportunity to make rare items available; and it is naturally this similarity to the distribution models of the industry that is seen as problematic from the film industry's viewpoint.

Where peer-to-peer file-sharing systems truly increase availability is through specialised sites that attract media-competent users. With users who are both media-literate consumers of film, television and games *as well as* computer literate, a reliable distribution works even with a low number of available files. In this case a great variety of formats is more a competitive advantage than a problem, because some specialists may prefer certain formats and file sizes. These are the kinds of sites defenders of file sharing most often refer to.

A SPECIAL CASE

Thus far I have focused on how peer-to-peer networks rely on different types of user behaviours and skills. One part of the problem for regular users is the different file formats. The files themselves tend to vary greatly, but have almost become standardised as either a full DVD (approx. 4.6 GB) or an '.avi' file small enough to be put on a CD (approx. 700 MB). Among the most popular downloads, full DVDs are quite rare. They require a very fast broadband connection, are sensitive to congestion (and also often cause congestion), and accordingly take a longer time to download. This certainly indicates that file sharing is not about copying DVDs but rather replacing them with smaller files, with less advanced image quality, better suited for the computer screen.

Of course, these highly compressed files change the quality of images and sounds. As this means that the actual digital object changes in form, one might ask if it indeed is the same film that is distributed, or if the variation in titling and technical quality also transforms the film as an object and artistic work.

There is a category of items that is very popular—particularly through sites without extra information—and is dominated by relatively small files, has very unreliable titling, and also (which is hardly an advantage) has a very unreliable relation between title and content: hardcore pornographic films.

Already the original retail format is different. The same episode (twenty to thirty minutes long) can appear in different combinations in different DVD collections, or as separate episodes online. Here, creative titling often determines the attraction and thereby affects the download frequency and behaviour. Following torrent tracker availability, it seems that title does not really matter—the topic and star names seem to be far more important. The flux of file identity helps create a high level of availability, even if most such files are normally unavailable. But this does not matter, since they are most likely available somewhere else under another title. The flipside of this is that it distorts statistics; not in data load, but there are definitely a great deal fewer porn titles available than first meets the eye.

I do not argue that the rest of the media industry should follow the porn industry (which also claims to be in crisis), but I do believe that looking at user behaviour concerning different types of material could help us more fully understand the phenomenon of file sharing. Obviously, something happens when files are smaller and the exact title cannot be established. On the other hand, if you cannot rely on a clearly defined title you need something else to identify the file. Therefore, perhaps the problem lies in the importance of identifying files. Commercially, clearly identified files are crucial, but as mentioned above it is often not the commercial object that is copied but rather a smaller version of it. The major exception is high-quality files of recent TV series episodes, which are not commercially available on DVD. Here, peer-to-peer networks seem to make up for TV broadcasters' delayed or non-existing distribution on other markets.

CONCLUSION

DVDs or CDs as commercial products have to rely on being collectable. The downloading of (reduced) files made from the most commercially viable DVDs and CDs seems to constitute a great part of the files distributed through peer-to-peer file-sharing systems (Martikainen 2010). Thus, the main driving force behind both formats is the idea of ownership. From this point of view, peer-to-peer file sharing has *not* changed any fundamental market mechanisms or market behaviours. File sharing is, in this respect, indicative of a disagreement on price.

The two forms of exceptions addressed here—the dissemination of particularly films and TV shows and episodes not available on any DVD or online market and the dissemination and 'repackaging' of porn films—open up for a more interesting discussion. In this respect peer-to-peer file-sharing systems become more interesting to study as examples of transitory systems, particularly regarding how media products become problematised in the meeting between the instability of the object, or product, and the relationship between object and value. In one sense this is an expression of user-*driven* (rather than user-generated) systems, that is, peer-to-peer as not only a technological but also an organisational feature (cf. Tapscott and Williams 2006; Jenkins 2006a; von Hippel 1986).

Looking at peer-to-peer file sharing this way, it seems as if users prefer a low image quality file for free, as well as a paid-for movie theatre ticket, to paying for a DVD. Based on what is distributed through peer-to-peer file-sharing systems, users/consumers seem to appreciate the production of short segments distributed as small files for free in ways that increase sales of movie theatre tickets, computer or video games, or TV series, or any other merchandise except for DVDs. For independent productions, or other productions unknown to the audience, unstable titling is a way to create interest or new brands. And as for the DVD, the problem is not primarily that it is copied but rather that its content is more attractive reduced in size and quality.

In the way peer-to-peer file-sharing systems work today there are certainly many buyers and many sellers, both willing and able to buy and sell at certain prices, but with a great degree of disagreement over the notion of value (and consequently of price) between them. The reason for this is partly that the consumer is already paying through his own work and equipment, partly due to a disagreement over what constitutes relevant information, and perhaps primarily due to the great variation in the products' characteristics.

Since all three aspects—transaction costs, information and product homogeneity—are fundamentally associated with how peer-to-peer file-sharing systems work, it is hardly surprising that no one has succeeded in creating a market system based on peer-to-peer file sharing.

It can therefore be concluded that both the representatives of the legal fight against file sharing and many of those in favour of 'free' culture place too much attention on controlling the system. Or, in Thomas W. Malone's words,

> If you are trying to control a system, you apply all your effort to making the outcomes of the system be what you want. If, on the other hand, you are trying to cultivate a system, you are not attached to the outcomes in the same way. (Malone 2004:181)

Such a cultivation of the system would in this case mean that the legal battle against file sharing, that is, the dissemination of low resolution video and audio files made from commercial DVDs or CDs, will either eliminate illegal file sharing or not do it. Both cases are typical results of market mechanisms. If this is done without eliminating the sharing of files that today have no commercial value, that is, those which are not available on DVD or CD or online, it is hardly a problem for the system—only for the potential sellers and buyers who disagree on the price.

What makes peer-to-peer file sharing interesting to study is the potential of what could happen if this system is cultivated rather that controlled. Then all 'peers' could gain access to an interesting tool for creating markets of audiovisual products, by discovering which products are popular enough to become commercial products. And those that do not become popular enough could without harm remain free, while simultaneously becoming accessible to the few who are interested in them.

What is needed, though, is a further analysis of the character and quality of the media object. Before this happens, peer-to-peer file sharing will continue to be an example of not only an extremely volatile and collaborative unstable market, but also a fertile ground for new film events.

REFERENCES

Anderson, C. 2006. *The Long Tail: Why the Future of Business Is Selling Less of More*. New York: Hyperion.

Babaioff, M., J. Chuang and M. Feldman. 2007. Incentives in Peer-to-Peer Systems. In *Algorithmic Game Theory*, edited by N. Nisan, T. Roughgarden, E. Tardos, and V. Vazirani. New York: Cambridge University Press, 593–610.

Balio, T. 1985. Part III: A Mature Oligopoly, 1930–1948. In *The American Film Industry* (rev. ed.), edited by Tino Balio. Madison: The University of Wisconsin Press, 253–400.

Boudet-Dalbin, S. 2007. When P2P Gets Work in Hollywood. *Intermedia: L'avenir de la distribution des films par Internet.* March 4. Retrieved from http://intermedia.homo-numericus.net/archives/42. Last accessed 2011–09–27.

Bruno, G. 1993. *Streetwalking on a Ruined Map: Cultural Theory and the City Films of Elvira Notari.* Princeton, NJ: Princeton University Press.

Charney, L. and V. R. Schwartz. Eds. 1995. *Cinema and the Invention of Modern Life.* Berkeley: University of California Press.

Deuze, M. 2007. *Media Work.* Cambridge: Polity.

Jenkins, H. 2006a. *Convergence Culture: Where Old and New Media Collide.* New York: New York University Press.

Jenkins, H. 2006b. *Fans, Bloggers, and Gamers: Exploring Participatory Culture.* New York: New York University Press.

Kaplan, E. A. Ed. 1990. *Psychoanalysis and Cinema.* London: Routledge.

Langer, F. 1913/2008. Cinema All the Time. In *Cinema All the Time: An Anthology of Czech Film Theory and Criticism, 1908–1939*, edited by J. Anděl and P. Szczepanik. Prague: National Film Archive, 92–97.

Machill, M., M. Beiler and M. Zenker. 2008. Search-Engine Research: A European-American Overview and Systemization of an Interdisciplinary and International Research Field. *Media, Culture & Society* 30(5), 591–606.

Malone, T. W. 2004. *The Future of Work: How the New Order of Business Will Shape Your Organization, Your Management Style, and Your Life.* Boston: Harvard Business School Press.

Manovich, L. 2001. *The Language of New Media.* Cambridge, MA: The MIT Press.

Martikainen, E. 2011. Does File-Sharing Reduce DVD Sales? Abstract available at SSRN: http://ssrn.com/abstract=1742443. Last accessed 2011–09–27.

McKee, A. 2007. Will BitTorrent Change Television? A Luddite's View. *FlowTV*, 6.04. Available at http://flowtv.org/2007/07/will-bittorrent-change-television-a-luddites-view/. Last retrieved 2011–09–27.

Moss, A. 2011. Piracy: Are We Being Conned?. *The Sidney Morning Herald*, March 22.

Ormerod, P. and R.A. Bentley. 2010. Modelling Creative Innovation. *Cultural Science*, 3(1), 1–15.

Papies, D. and M. Clement. 2008. Adoption of New Movie Distribution Services on the Internet. *Journal of Media Economics* 21(3), 131–157.

Sanchez, J. 2008. 750,000 Lost Jobs? The Dodgy Digits Behind the War on Piracy. *Ars Technica*, October 7.

Tapscott, D. and A.D. Williams. 2006. *Wikinomics: How Mass Collaboration Changes Everything.* New York: Portfolio.

Vaidhyanathan, S. 2011. *The Googlization of Everything (and Why We Should Worry).* Berkeley: University of California Press.

von Hippel, E. 1986. Lead Users: A Source of Novel Product Concepts. *Management Science*, 32(7), 791–805.

5 Technostruggles and the Satellite Dish

A Populist Approach to Infrastructure

Lisa Parks

Most people are socialised to know very little about the infrastructures that surround them in everyday life, whether electrical systems, sewer pipes or broadcast networks. Not only are people socialised to be unaware of such systems; infrastructures are often designed purposefully to be invisible or transparent, integrated with the built environment, whether submerged underground, covered by ceilings and walls, or camouflaged as 'nature'. Further, since infrastructures often extend across vast territories they are impossible to grasp in their entirety and are difficult to describe. Much has been written about the social history of transportation and communication infrastructures. Studies by Wolfgang Schivelbush, Harold Innis, James Carey and others have helped us to understand the railway, telegraphy and broadcasting in relation to structures of industrialisation and modernity. Insisting upon the relevance of culture to such systems, Carey writes, 'Technology, the hardest of material artifacts, is thoroughly cultural from the outset: an expression and creation of the very outlooks and aspirations we pretend it merely demonstrates' (Carey 1992: 9). This chapter explores the topic of infrastructure from a cultural perspective using a partial, object-centred, and knowledge-oriented approach. Here we might access the work of cultural studies scholar John Fiske and consider what it might mean to study infrastructure from a perspective of 'popular knowledges' and 'technostruggles'.[1] As Fiske reminds us, 'The multiplication of communication and information technologies extends the terrains of struggle, modifies the forms struggle can take, and makes it even more imperative that people grasp the opportunities for struggle that the multiplying of technologies offers' (Fiske 1996: 240).

A populist approach to infrastructure emphasises how people come to access, imagine and understand infrastructures, not only by demystifying their seamless operation in everyday life, but also by understanding them as material forms—sites, nodes, parts, pieces, objects to be seen, engaged with, handled, felt and investigated. To explore the possibility of a populist approach to infrastructure, I focus on the satellite dish, an object perched upon millions of residential and commercial buildings in urban and rural settings around the world. When direct satellite broadcasting emerged in

Europe and Asia during the late 1980s and early 1990s, Charlotte Brunsdon published a pathbreaking essay called 'Landscapes of Taste' in which she analysed how the satellite dish had been discursively positioned as a signifier of bad cultural taste in London as it made private choices about television viewing public. While these notions of bad taste may linger, the satellite dish has taken on a wider range of meanings with the worldwide growth of satellite television services over the past thirty years.

Today the satellite dish not only indexes class and taste formations; as we shall see, it also has been used to single out immigrant ethnic communities in urban space. In addition, the satellite dish has also been used as a decorative surface, as a technology of local control, as a repurposed object and as a residual material. Positioned at the edges of satellite infrastructure, the dish is the peoples' portal to a much broader, more expansive and complex global assemblage that is predicated upon a particular organisation of resources and spaces ranging from heavy metal deposits to building facades, from bandwidth in the electromagnetic spectrum to slots in the geostationary orbit. Since it would be impossible to visualise the world's entire satellite infrastructure in a single frame or view, we might think of the satellite dish as the most common and hyper-visible part of this infrastructure. As such it can be used to encourage inferential imaginings of a broader assemblage of technologies, earthly sites and orbital domains, institutional structures and capital accumulations of which the satellite dish is just one small part.

In the first section of the chapter I consider what is at stake in infrastructure concealment or transparency and explore how infrastructures become intelligible to people in everyday life. I delineate five modes of infrastructural intelligibility and describe how people can become aware of infrastructures through their encounters and experiences with different infrastructure parts, sites, stages or uses. In the second section, I focus on the satellite dish, a highly visible part of satellite infrastructure with which most people are familiar. Situated in urban and rural locations around the world, the satellite dish not only connects viewers to television programming, its installation and use brings a series of power plays into relief. The satellite dish is used to blend worlds, rattle totalitarian regimes, link to homelands, profile and monitor ethnic communities and imagine the broader satellite system of which it is part. As I try to demonstrate, this seemingly most banal of objects is tethered to the most serious kind of social and political struggles.

INFRASTRUCTURAL INTELLIGIBILITY

Freeways, oil pipelines, sewer systems and television networks are just a few of the infrastructures that shape everyday life in industrialised societies. While some infrastructure sites are celebrated as sites of technological spectacle and wonder, others are hidden beyond thick green shrubs or buried deep underground. Television towers from Berlin to Tokyo, from Stockholm to Tehran, were designed precisely to be seen and admired from near and far. Jutting hundreds of feet into the sky, TV towers are capped with visitors' centres, restaurants/bars and viewing platforms, encouraging people to tour these infrastructure sites and glimpse what the world looks like from a transmitter's perspective. Other infrastructure sites such as transoceanic cable landings, however, are patently off limits to the public. Not only is access to such sites forbidden; violations are often deemed to breach national security. During research visits to cable landings throughout the Pacific, media scholar Nicole Starosielski (forthcoming) encountered warning signs, fences and security systems obstructing access to the sites, arguing the cables are protected by 'strategies of insulation'. Finally, cell phone towers are often purposefully covered up with plastic disguises, whether as a pine tree, a minaret or an osprey nest, so that they can blend in with the local environs (Parks 2009). Such 'concealment strategies', as they are known, represent a concerted effort to make infrastructure sites imperceptible.

As these brief examples suggest, an infrastructure can be differentially positioned on a continuum of visibility and invisibility depending on its material composition, scale, design, location and purpose. While some infrastructure sites are purposefully concealed so that they blend in with the urban or natural environment, others become transparent over time as they become part of the seamless patterning of everyday life. Stephen Graham and Simon Marvin refer to this tension between visibility and invisibility as a 'dialectics of invisibility and monumentalism' (Graham and Marvin 2001: 20), suggesting that when infrastructures are not 'being hidden' they are 'revealed, celebrated and constructed a iconic urban landmarks as embodiments of the "phantasmagoria" of particular urban times and places'. In his *Field Guide to Infrastructure*, Brian Hayes offers photographs and descriptions of the 'power lines, water tanks, streetlights, manholes, traffic signals, cellular towers—that we pass by every day and yet seldom really notice', and suggests, 'These are places that most of us never see close up; many of us would go out of our way to *avoid* seeing them' (Hayes 2005: 1).

The transparency and concealment of infrastructures has several possible effects. First, the inability to perceive infrastructures inhibits opportunities for technological literacy among citizen-consumers. It prevents individuals from recognising and inquiring about the material extensions they use, subsidise and are surrounded by in daily life. Second, and related to the first point, infrastructural concealment can cultivate and reinforce hierarchies of knowledge where those who own, see and control infrastructures have the power to understand them and those who subsidise and use them know very little, if anything, about them. Third, transparency and concealment keeps citizen-consumers out of the loop with respect to the design of and creative vision for infrastructure sites, even though many are located in or near public spaces. When citizens have been involved in infrastructure site design, they have turned electrical exchange boxes into vibrant murals, train station walls into tiled mosaics, and telecommunication transponders into massive paintings, intensifying visual activity around infrastructure sites and generating further interest in them. Finally, if infrastructure sites are to be placed in the built environment near residences and workplaces, citizens have a right to know what they are made of, how they function and what the long-terms effects of their operation might be upon public health and the natural environment.

The intelligibility of infrastructures is interwoven with the politics of knowledge, place and aesthetics. To provide a sense of the ways that citizen-consumers encounter and experience infrastructures in everyday life, I briefly describe five modes of infrastructural intelligibility: 1) design and planning, 2) construction zones, 3) infrastructure sites, 4) system failures and 5) portal technologies. By engaging with infrastructure in this way, I am interested in a phenomenology of infrastructure—an attempt to understand the moments of life and conditions in which individuals come to realise and know that infrastructures exist. *Infrastructural intelligibility* can thus be understood as the process by which ordinary people use observations, images, information and technological experiences to infer or imagine the existence, shape or form of an extensive and dispersed infrastructure that cannot be physically observed by one person in its entirety.

Design and Planning

Typically designed on paper or on screen using computer-aided design (CAD) programs, infrastructures are visually represented in schematic designs or blueprints as flow charts with specialised symbols representing interlinked nodes. Though such representations usually circulate among a community of experts, they are also presented at public hearings and sometimes published in newspapers so that citizen-consumers can comment on the development of an infrastructure during its planning stages. While reading schematic designs requires specialised skills that many citizen-consumers do not possess, such representations can be used to activate a series of questions or an interrogative disposition towards infrastructure during the planning process. This is especially important if there are concerns about the environmental or public health impacts of its installation.

Construction Zones

A second mode of infrastructural intelligibility involves physical observation of construction zones. Infrastructure development often involves extensive earth movement such as trench digging as well as the transportation of large components, whether miles of cable or huge pipes, across freeways, railways or shipping routes. Since infrastructure projects take years to complete, people must take detours around construction zones and cannot help but notice them. In such conditions, consumer-citizens can become aware of infrastructure, recognising it as an evolving assemblage of pieces and parts. After construction, many infrastructure sites are covered with earth never to be seen in public again. Construction thus offers a twilight of visibility, revealing infrastructures as skeletal structures that will eventually become invisible, whether buried underground or concealed behind walls.

Infrastructure Sites

A third mode of infrastructural intelligibility occurs through the observation of individual infrastructure sites, whether a television tower, a power line or a shipping port. Such sites are not only interesting in their own right, but can also trigger inferential imaginings of the larger system of which they are part. Projects such as Brian Hayes' (2005) *Field Guide to Infrastructure*, Kazys Varnelis' (2009) *The Infrastructural City* and Ursula Biemann's (2005) *Black Sea Files* stimulate critical reflection upon infrastructure sites such as nuclear power plants in the US, cell phone systems in southern California and oil pipelines in central Asia. Using photography, video, maps and commentaries, their work isolates infrastructure sites implicated in the distribution and consumption of energy, signals and oil and draws attention to vital dimensions of the material world often overlooked. Their work demonstrates that a focus on one infrastructure site, whether through direct observation in the field or audiovisual representation, foregrounds the unique scale and challenge of infrastructural representation.

System Failure

Infrastructural intelligibility also emerges during moments of system failure. The scale of the failure can be large or small, internal or external, rapid or extended. When the lights go out, e-mail messages are not sent, or the drain clogs citizen-consumers become aware of their relation to and reliance upon a broader system that they cannot see and (most) do not know about. Moments of failure reveal that infrastructures are not inert systems, but rather are intricately interwoven with psychic investments, labour, creativity, movement, communication and other everyday flows. Infrastructures are affective systems in the sense that they both affect life conditions and generate affective responses. Infrastructures become most visible when they fail because their failure can proliferate and cascade into a series of other failures. For instance, when the lights go out, work may not get done, a paycheck does not appear and food is not put on the table. Or when a solar flare causes a satellite to malfunction, ATMs do not deliver cash, a taxi ride is not possible, a crucial meeting is missed and a deal is lost. As Leigh Star observes, breakdowns become 'the basis for a much more detailed understanding of the relational nature of infrastructure' (Star 1999: 382). While failures bring the relationality of infrastructures into relief, most citizen-consumers do not have the knowledge, skills or authority to fix them. Infrastructure failures thus not only generate other failures in everyday life; they are symptomatic of a broader social failure: the centralisation and hierarchisation of technical knowledge.

Portal Technologies

Finally, infrastructures become intelligible during use. Infrastructural connection becomes meaningful the moment the tires hit the road, the dishwater disappears down the drain, the mobile phone rings or the television program appears on the monitor. Infrastructure use is mediated by consumer technologies such as the automobile, the kitchen sink, the cell phone and the satellite dish. These physical objects enable consumer access to and use of roadways, sewer pipelines, cellular phone networks and satellite TV systems and as such are vital to infrastructural operations. Because of this they can be understood as *portal technologies* that enable and structure users' infrastructural connections. Their everyday handling of and reliance upon such portal technologies means that citizen-consumers tend to be more psychically invested in these tactile objects than in the extensive infrastructures to which these objects are linked. Most people do not have deep knowledge of the way their cell phone network is organised, but they do know how to use their cell phone to connect to it. In this sense, these objects can be understood as affective stand-ins or physical manifestations of massive, dispersed systems that cannot be seen or touched.

If most consumer-citizens experience infrastructures through the objects or portal technologies that they use on a daily basis, then struggles over infrastructural futures take place not only within the official spheres of urban design and planning, but also within cultures of everyday technological use. A populist approach to infrastructure involves recognising the multiple ways that citizen-consumers use portal technologies as part of the process of accessing, understanding and negotiating infrastructures. These everyday objects are crucial to popular knowledge formations about infrastructure. As Nigel Thrift (2000: 380ff) suggests, human agency emerges through technosocial assemblages that are shaped by the 'little things' of practice rather than the grand visions of technophiliac fantasy.

With this in mind I would like to turn to a focused discussion of a 'little thing' known as the satellite dish. As a portal technology for direct satellite broadcast infrastructure, the satellite dish connects its users to a circuitous assemblage of satellites, orbital paths and earth stations. As one of the most familiar technological objects in urban and rural areas around the world, the satellite dish can be used both to downlink television programming and to conjecture a vastly dispersed and largely invisible satellite infrastructure. As an object of attention, a surface of inscription, a site of value and a coveted good, the satellite dish localises and materialises social struggles and is imbricated in the politics of expression, place and planetary control.

SATELLITE DISH MATTERS

Satellite dishes are manufactured in sites around the world, with industry leaders in China and Taiwan, and firms generate millions of dollars in profit each year in the process of making and selling them. Most satellite dishes are made from materials such as aluminium, steel, copper, fibreglass and plastic. They are traded in urban and rural marketplaces and are part of formal and informal economies. While there is a robust global satellite dish industry, the making and installation of satellite dishes varies from place to place. At the black market in Ulaanbaatar, Mongolia, parts are imported from China and some are repackaged and sold under the Malchin label (*malchin* means 'cowboy' in Mongolian language), and then are taken to remote locations, assembled, and fuelled with solar and wind power (Figure 5.1). In Istanbul an entire street, Yusek Kaldirim Cadessi in Beyoglu, is lined with satellite dish dealers. The dealers import dishes from China, brand them with names like Dreambox, NextStar and Wavefield, and mount them outside their shops along the street to attract customers (Figure 5.2). In Thailand, grade-school children devised a system for making satellite dishes out of salvaged aluminium pot and pan covers. After receiving donations of old pots and pans from villagers, the kids hammered them into a parabolic shape, assembled stands from steel and starting selling them around the country. The project has been implemented in 300 Thai schools as a source of income for the schools (Saterangsee 2010). And anyone with Internet access can find a YouTube video that demonstrates how to make a satellite dish out of a kitchen colander used to drain pasta.[2]

Whether mass-produced or handmade, the satellite dish is a key part of an expansive infrastructure used for the global trafficking of audiovisual signals. The dish consists of a parabolic or bowl-shaped metallic surface and a feed horn. Its surface captures electromagnetic waves transmitted from a given satellite and reflects them onto the feedhorn. Inside the feedhorn a low noise block (LBN) converter amplifies the signal bouncing off the dish, filters out the noise and sends the signal through a cable to a satellite receiver box and then to a television monitor. The satellite dish mediates activity between a user on earth and a satellite in orbit. Often physically installed and adjusted by citizen-consumers, it is portal technology that facilitates access to and knowledge of a broader satellite infrastructure. It is a tactile connection to and a physical stand-in for an orbital system that is unintelligible to most.

While satellites are relatively invisible, satellite dishes are hyper-visible. They can be seen in urban and rural, wealthy and poor, central and remote areas around the world. Large apartment buildings in cities such as Berlin, Fez, Istanbul, Calcutta and Amsterdam are blanketed with satellite dishes. Satellite dishes appear on building facades, balconies and rooftops, and are surrounded by plants, drying laundry, solar panels, water heaters and patio

Figure 5.1 Satellite dishes and solar panels are sold at the black market in Ulaanbaatar, Mongolia. Photo by author.

Figure 5.2 Satellite dish vendors place dishes in front of shops along the street in Istanbul. Photo by author.

Figure 5.3 Satellite dishes on an apartment building in Manipur, India. Photo by Bhaskar Sarkar. Used with permission.

furniture. While some satellite dishes are shiny and new, others are covered in rust or bird droppings. They come in different sizes, face different directions and receive signals from different satellites. If an apocalypse were to befall the planet tomorrow, survivors would find millions upon millions of satellite dishes amidst the ruins.

While satellite dishes are relatively easy to purchase, assemble and see, there are often complex politics surrounding their installation and use, and these politics vary from place to place. Some property managers patently forbid their installation for aesthetic reasons. This is especially an issue in relation to historic buildings or upscale architecture. In 2004 the Münster-based architecture firm Bolles-Wilson reversed this trend by explicitly integrating satellite dishes and air conditioners into a paint design they were commissioned to develop for a communist style apartment building in Tirana, Albania. They called the concept a 'Marriage of Air Conditioners and Satellite Dishes' and used white squares on a red and orange striped building to mimic and simulate the random positioning of air conditioners and satellite dishes already on the façade (Zacks 2005). Rather than eliminate these 'awkward appendages', the design conceptualises the satellite dish and the air conditioner as fundamental to post-Communist visions of the built environment, linking cool air and global television to a life of comfort, mobility and leisure.

In other parts of the world, satellite dishes sit on or next to buildings that are nearly falling down. For instance, in the California desert, many remote and barely standing homesteads are equipped with satellite dishes.[3] David Morley (2000: 282) points out that real estate agents often see the satellite dish as a sign that property values are going 'down market' and that house

Figure 5.4 Satellite dish on a home in the desert town of Palo Verde, California. Photo by author.

prices in the area are about to fall. In this context, however, the real estate market is already severely depressed and the satellite dish is simply a sign of human life. My trips to the desert communities of Palo Verde and Yucca Valley (Parks 2007) in recent years revealed many dilapidated homes with at least one, if not two or three, satellite dishes mounted or strewn about the yard (Figure 5.4). Many owners of low-income desert households do not invest in exterior paint jobs or structural upgrades, but they do invest in satellite television. Some of these residences are 'off the grid', using local water and energy supplies, yet are fully integrated within the global satellite television system. While some residents of these communities are elderly people without pensions or working poor, others live in the desert simply because they want to be left alone. In such remote areas, satellite dishes become highly valued and coveted objects because they facilitate symbolic integration while preserving geophysical separation.

To reinforce the important role of the satellite dish in relation to low-income dwellings around the world, Slovenian artist Marjetica Potrč often integrates satellite dishes into her multimedia installations. Potrč's installations are inspired by low-income homes or municipal 'core units' (structures that include basic access to water and sewage) that she has encountered during her collaborations in cities, islands and remote communities around the world. All of the installations feature a source of energy and a mode of communication. For instance, her installation 'Forest Rising' is a rural school equipped with a satellite dish and solar panels, perched on island elevated by tree trunks. As Potrč explains, 'The project is based on practices developed by Amazonian communities in Brazil in response to the most pressing social, economic, and environmental concerns of the 21st century', including 'the protection of knowledge, the protection of territories, and global connectivity'.[4] Spotlighting the satellite dish's reliance upon energy infrastructure (in this case, solar panels), Potrč also emphasises its use as a technology of mobility and survival by low-income people around the world. Her work reveals that the presence of a satellite dish can make even the tiniest makeshift dwelling appear to be 'lived in' or 'occupied'.

During the last decade, the visibility of satellite dishes in Europe has intersected increasingly with anti-immigration politics and anxieties about terrorist threats, and the satellite dish—that everyday object that demarcates a 'peoples' link' to the global satellite infrastructure—has taken on a more volatile status in the context of the war on terror. In his short essay 'Squaring the Dish', architect Mark Collins critiques Dutch conservatives' efforts to deprive immigrant communities of satellite television programming from North Africa and the Middle East, whether by pushing property owners to sue immigrant residents for installing dishes or threatening to demolish the apartment buildings that they inhabit.[5] Communities on the periphery of Amsterdam's centre are sometimes referred to as 'dish cities' or 'satellite cities' because of the many satellite dishes that immigrants from

Turkey or Morocco have installed to receive television signals from North Africa and the Middle East (Buruma 2005).

Such 'dish cities' have been the subject of attack among political conservatives. In an article titled 'Europe's Angry Muslims', Robert Leiken points out that Mohammed Bouyeri, the Moroccan immigrant that assassinated Dutch filmmaker Theo Van Gogh in 2004, 'grew up in a proletarian area of Amsterdam sometimes known as Satellite City because of the many reception dishes that sit on its balconies, tuned to al Jazeera and Moroccan television', implying that satellite television was somehow connected to the murder (Leiken 2005). Geert Wilders, a member of the Dutch parliament and chairman of the Party for Freedom, gave a speech in New York in September 2008 to warn Americans about the 'Islamicization of Europe'. In it he mentioned that '[s]atellite dishes are not pointed to local TV stations, but to stations in the country of origin' and went on to claim, 'They [referring to Muslim immigrants] do not come to integrate into our societies; they come to integrate our society into their Dar-al-Islam. Therefore, they are settlers' (Bostom 2008). This same logic is being applied to other 'dish cities' in Berlin, Brussels, Milan, Madrid and Paris as well. Even before 9/11 Hargreaves and Mahdjoub (1997) revealed in their study of satellite television viewing among ethnic minorities in France that politicians had become anxious about 'the idea of satellite television beaming fundamentalist propaganda into the *banlieus*', explaining, 'Many . . . have seen in these channels at the very least a threat to the cultural integrity of France, as well as a potential threat to the country's social and political stability' (Hargreaves and Mahdjoub 1997: 461).

The fixation on the 'dish cities' and the satellite dish in the discourse of political conservatives is telling. In addition to exposing anxieties about the transnational distribution of television signals from North Africa and the Middle East, this discourse suggests just how powerful and significant infrastructural visibility can be. While immigrant communities in Europe also use the Internet and mobile telephony to communicate across national borders, laptops and cell phones do not have a constantly visible reception apparatus and can be more readily concealed. Politicians have tended to focus upon the satellite dish because of its public presence and visibility. A scattering of satellite dishes on a building facade is seen in some European cities as a visual correlate for the presence immigrant families even though many white Europeans subscribe to satellite television as well. In this case, systems of ethnic differentiation are based not only upon skin colour, language or birthplace; *ethnic communities are being differentiated and identified based on the technologies they consume and the infrastructures to which they connect.* In some urban spaces the satellite dish itself has become a marker of ethnic identity. As a result, the satellite dish is increasingly implicated within practices of ethnic profiling and surveillance within urban spaces.

Indeed, satellite dishes not only link diasporic communities to homeland television signals, they have increasingly been used to reinforce a 'class of civilisations' ethos and to rationalise post-911 global security paradigms. In this context, simple acts of satellite dish purchase and installation (not to mention television viewing) can position immigrants as 'suspects'. In an analysis of Amsterdam's Slotervaart neighbourhood, one of the 'dish cities,' urban studies experts Ralf Brand and Jon Coffee observe,

> The line between paranoia, surveillance, safety, stereotypes and coexistence is . . . blurred by such seemingly mundane artefacts as satellite dishes or net curtains. The former are often taken by Dutch people as evidence of a high presence of immigrants trying to stay in touch with their country of origin. Net curtains are often interpreted as an attempt to 'hide something'.[6]

The mere presence of satellite dishes or certain types of window coverings, in other words, are read by many white Europeans as evidence of potentially dangerous activity. As David Morley writes, 'In some parts of Europe, they [satellite dishes] have come to symbolise forms of "cultural treason" against their "host nation"' (Morley 2000: 282).

In a discussion of the London-based Kurdish satellite television network MED-TV, which aired from 1995 to 1999, Monroe Price explains that the physical position or attitude of the satellite dish became a way of distinguishing Kurdish from Turkish residents in Turkey. As Price explains,

> for viewers to turn their receiving dishes to Hotbird meant they were conspicuous in comparison to their neighbors, whose dishes were turned to a different, Eutelsat-oriented satellite, which carried traditional Turkish entertainment channel services. The difference in the attitude to the dish could be perceived by the authorities. To protect its viewers, MED-TV had to shift, therefore, from Hotbird to Eutelsat. (Price 2002: 156)

MED-TV was eventually shut down, due in part to pressure from the Turkish government, who claimed the satellite network was the media arm of the PKK, and because the satellite network was deemed to violate the British Independent Television Commission's standards of objectivity and impartiality (Price 2002; cf. Sakr 2002: 61ff). This case reveals that both the visibility and the position of the satellite dish have been used to make assumptions about residents' ethnicity and political agendas.

The satellite dish has become the subject of controversy in post-communist countries as well. In 2007 Turkmenistan's President, Kurbanguly Berdymukhammedov, demanded that 'ugly' satellite dishes be removed from all homes in the capital of Ashbagbat. At that time 600,000 residents of the city depended on satellite dishes to receive television signals from

Russia, Turkey and Arab countries.[7] Berdymukhammedov later indicated that the satellite dish ban was 'intended to remove a blight on the skyline and make Ashgabat a prettier city' and claimed that the scattering of dishes would be replaced with 'a single powerful dish on each building' (Saidazimova 2007). Berdymukhammedov reiterated his position yet again in August 2011, ordering the dismantling of all satellite dishes owned by citizens of Turkmenistan. Some felt this was a state response to international satellite news coverage of an arms depot explosion that had occurred just outside Ashgabat in Abadan on July 7, 2011. The event went uncovered by Turkmenistan news agencies, but appeared in satellite news broadcasts after reports from Turkmen citizen journalists and bloggers travelled on the Internet and were picked up abroad. The satellite dish became part of an intermediale feedback loop, where a newsworthy event in Turkmenistan suppressed by the state broadcaster was transmitted via cell phones and blogged about via Internet and then relayed back into Turkmenistan via international satellite television networks. Some perceive the satellite dish ban as part of a state-led media containment strategy, particularly since many of those who blogged about the arms depot explosion have been apprehended and interrogated (Kilner 2011).

One Turkmen journalist, Serdar Aytac, insists 'the very appearance of these [satellite] antennas are nothing more than a response of society to the authorities'.[8] Turkmen blogger Tomyris disputes the statement that satellites are 'ugly', claiming,

> Satellite dishes in Turkmenistan are as beautiful to the media guru's eye as a mountain is to the hiker's. There is something astoundingly enriching about seeing individual satellite dishes as symbols of education—of the transmission of flowing information—and one's ability to be connected to the world around them. The dishes are symbols of information—news and entertainment—being brought inside and outside of a country where the media itself has hardly any rights.[9]

Similar satellite dish banning and demolition has occurred in China, Tibet and Iran in recent years.[10] And, of course, intermediale feedback loops also shaped the media coverage of the Iranian revolution of 2009 and the historic revolutions of the Arab spring in Tunisia, Egypt and Libya in 2011. Thus while many find the satellite dish to be a blight on the landscape or irrelevant in the age of the Internet, citizens in different parts of the world continue to insist upon its usefulness and political significance.

The situations of MED-TV and Turkmenistan serve as reminders that satellite dish access has been protected as a human right in Europe since 1990. A 2011 decision in the European Court of Human Rights in Strasbourg involving two tenants from Sweden, who were evicted from their apartment after installing a satellite dish, confirmed that having a satellite dish is a human right under the Freedom of Information Act (Drury 2011).

The court ruled that banning satellite dishes on listed buildings, social housing or private homes could breach citizens' right to free expression and reinforced a 1990 ruling that indicated 'reception of foreign satellite TV channels via satellite dish or other antenna' is protected by Article 10 of the European Convention on Human Rights (Briel 2011). Using the occasion as an opportunity to promote Europe's free market economy, the European Commission's Internal Market Commissioner Frits Bolkestein responded to the ruling by proclaiming,

> The right to use a satellite dish [is] one of the many concrete benefits for European consumers of the free movement of goods and services within the internal market. Satellite dishes are an increasingly popular tool for receiving multiple services via satellite: they facilitate mutual exchanges between our various cultures by overcoming national borders, and familiarise the general public with the new remote communications technologies. Their use must therefore be free from any unjustified obstacle. (Drury 2011)

This statement not only articulates a liberal pluralist ideology that is fully at odds with the unequal treatment of satellite dish owners in recent years; it also reveals how tightly tethered the satellite dish is to neoliberal visions of an integrated European marketplace.

Despite the legal protections mentioned already, the condition of satellite dish–based suspicions has led some neighbourhoods in Europe to respond expressively by 'pimping out' their satellite dishes with colourful paintings and designs. In 2007 a Dutch artist led an initiative to work with children in one of Amsterdam's 'satellite cities' inhabited primarily by first- and second-generation immigrants.[11] Here the surface of the dish is treated as a site of inscription and distinction. While this act may have drawn even further attention to this 'dished' building, it also enabled communities inhabiting the building to symbolically respond to and attempt to counteract the anxieties of conservatives. Indeed, this 'psychedelic dish movement', as one blogger described it, was part of an attempt by immigrant communities to mitigate the perceived threat of their downlinking of television signals from North Africa or the Middle East in the context of the global war on terror.[12]

This move to claim the surface of the satellite dish as a space of local control and inscription flagrantly displays what a populist approach to infrastructure might look like. Peoples' control over the dish itself, whether through the act of making it, positioning it, using it or decorating it, is significant because it reveals the multiple, expressive and contentious ways in which citizen-consumers connect with satellite infrastructures in the personalised spheres of daily life. Such acts have even resulted in an Amsterdam-based spinoff business titled Satellitedishsticker.com where one can buy decorative stickers for 35–45 Euros to place on the dish, whether in the form of a kitten, a sunflower, a Turkish pattern, a European Union flag or a peace sign. Given the symbols available, Satellitedishsticker.com has clearly capitalised on the satellite dish controversies in Europe, encouraging dish owners to explicitly make a statement, whether to express national allegiance, call for world peace or simply celebrate kitsch.

While some may assert infrastructural control by ornamenting the dish, other individuals have devised different ways of dealing with the infrastructure's edge, mounting so many dishes that it would be altogether impossible to determine the resident's ethnicity or political agenda. Al Jessup, a former ice cream salesman who is disabled, positioned twelve dishes on the balcony of his small home in Bethany, West Virginia. He claims his fascination with satellites changed his life and his house. 'Up in the sky—there's lots of free stuff', he says. Jessup positioned ten of his twelve dishes to receive 'free to air' signals from satellites such as Galaxy 10, AMC 2 and Telstar 5, and uses his dishes to find local news signals from states in the US where he used to live, including Tennessee and Louisiana (Figure 5.5) (Pridemore 2005).

On the other side of the world, South Korean, Lee Si-kap is nationally known as 'antenna man'. The farmer who tends pepper and sesame fields by day also owns and operates eighty-five satellite dishes from his home and uses them to downlink 1,500 channels from more than 100 countries (Sang-Hun 2009). A reporter indicates, 'When Mr. Lee looks skyward, he says, he imagines satellites in orbit. To him, the air is filled with broadcast signals, "like seeds from thistles"' (Sang-Hun 2009). Both Lee Si-kap and Al Jessup have arranged and used their satellite dishes in ways that allow

Figure 5.5 Al Jessup adjusting a satellite dish at his home in Bethany, West Virginia. Reproduced with the kind permission of *The Register-Herald*, Beckley, WV.

Figure 5.6 Satellite television dishes cover the home of Lee Si-kap, a farmer. Mr. Lee owns eighty-five dishes, the most of any South Korean. Photo by Choe Sang-Hun, reproduced with kind permission of *NY Times/International Herald Tribune.*

them to imagine and infer the presence of a broader infrastructure as well as establish personal links to it. The multiple ways that people engage with satellite dishes is significant not only because it complicates assumptions about infrastructural control and uniformity (since we see lots of variation at the edges); it is also suggestive of a peoples' interest in the physicality and physics of media infrastructures.

CONCLUSION

The satellite dish remains deeply imbricated within 'landscapes of taste', as Brunsdon suggested twenty years ago, and it is increasingly bound up in a broader set of technostruggles that extend from the making of dishes to architectural aesthetics, from property values to anti-immigration politics. Taking this portal technology seriously involves recognising the various ways that this object is implicated within concerns that extend from the local and everyday to the global and the geopolitical. In closing, I want to make three points. First, *infrastructure* is the word used to refer to a large-scale, spatially dispersed system that is controlled centrally, but may support a variety of technologised practices. What the preceding examples suggest, is that infrastructures are not uniform throughout and may be differentiated or reinvented at their edges or fringes, at the sites where people figure out how to make them part of their everyday lives. People make, decorate, collect or repurpose satellite dishes for a variety of reasons. These creative practices may flourish in proportion to the level of detachment and alienation that most people feel when faced with satellite television's technicalities. People may not be able to control and shape the programming that enters their homes, but they can make, position, use and repurpose the object that brings signals into their spaces, whether the school, the desert homestead, the city apartment building or the farm.

Second, the visibility of satellite dishes can play a role in allowing people to infer or imagine the materiality and significance of media infrastructures and processes that are imperceptible to us, but that play a key role in shaping our everyday experiences. The materiality of the dish always raises questions about where signals are coming from. A populist approach to infrastructure calls for a form of criticism that is organised not upon the spectacle of the monumental or the new, but upon the materiality of the thing—a form of criticism that leads to investigations of the raw materials, resources, labour, affect and energy required to sustain a system of satellite television in the first place as well as to its hidden (and often contested) cartographies. As Bruno Latour writes, 'Each object triggers new occasions to passionately differ and dispute . . . Objects . . . bind all of us in ways that map out a public space and all of its attendant stratifications' (Latour and Weibel 2005: 15). To be sure, the satellite dish has been at the heart of impassioned differences and disputes, and its use has generated newly stratified spaces in relation to the war on terror and post-communism.

Finally, studying the satellite dish can bring the politics of infrastructural intelligibility into relief. To develop an awareness of and interest in infrastructure, it is not necessary to be an operator, owner, engineer or regulator. The most familiar and tactile part of a broader invisible system can become a crucial catalyst of curiosities, investigations and experimentations. While infrastructures have their own spatial and temporal limits, they can be conceptualised and studied in many different ways. We do not know how long the satellite dish or the satellite economy will be in use, but we do know that there is an enormous capital investment in an industry and infrastructure that extends from satellites in orbit to earth stations on the ground, and from television studios to satellite launch pads. Despite the fetishization of handheld digital devices, whether laptops, mobile phones or e-readers, billions of people in the world still use satellite dishes. It is my hope that we will continue to approach the satellite dish and other media infrastructures as platforms for thinking through key issues of technological literacy, social power and resource use.

NOTES

1. A shorter version of this paper was presented at the Fiske Matters Conference in at the University of Wisconsin–Madison in June 2010, and is inspired by the scholarship of John Fiske. The shorter version was also presented at an HMKV workshop titled 'Satellite/Border/Footprint' in August 2010.
2. 'DIY Satellite TV: You're your Own Satellite Dish', posted by ITRTECH-FILMS, October 13, 2008, available at http://www.youtube.com/watch?v=YgLfP_S3iQE&feature=fvsr. Last accessed 2010–05–10.
3. For a longer discussion of satellite television in the California desert see Lisa Parks (2007).
4. Marjetica Potrč, 'Contemporary Building Strategies: Forest Rising', Marjetica Potrč website, available at http://www.potrc.org/project1.htm. Last accessed 2011–09–01.
5. Mark Collins, 'Squaring the Dish', *Slumlab Newsletter*, Spring 2009, available at http://www.proxyarch.com/media.html. Last accessed 2010–05–26.
6. Ralf Brand and Jon Coffee, 'The Urban Environment: Mirror and Mediator of Radicalisation', project summary available at http://www.urbanpolarisation.org/index.php?option=com_content&view=section&layout=blog&id=7&Itemid=81&lang=en. Last accessed 2010–05–26.
7. 'Turkmen Leaders Bar "Ugly" Satellite Dishes', *Reuters*, UK edition, December 4, 2007, http://uk.reuters.com/article/2007/12/04/tech-turkmenistan-satellite-dc-idUKL0472213120071204. Last accessed 2011–09–01.
8. Serdar Aytac, 'Turkmenistan: Another Assault on the Freedom of . . . ', blog of radio station Эхо Москвы, August 16, 2011, http://echo.msk.ru/blog/aytakov/802974-echo/. Last accessed 2011–09–01.
9. Tomyris, 'Make Turkmentistan Prettier—Rid One Satellite Dish at a Time', *New Eurasia blog*, August 20, 2011, http://www.neweurasia.net/media-and-internet/make-turkmenistan-prettier-%E2%80%93-rid-one-satellite-dish-at-a-time/. Last accessed 2011–09–01.
10. Lhumbum Tashi, 'Tibetan TV Dishes Pulled', *Radio Free Asia*, http://www.rfa.org/english/news/tibet/Tibetandishes-06202009092817.html. Last accessed 2011–09–02; 'Iranian Police Dismantle Satellite Dishes', *News.Az*, August 11, 2011, http://test.news.az/articles/iran/42292. Last accessed 2011–09–06; 'Police in Iran Destroy Satellite Dishes', *Advancing Human Rights blog*, August 18, 2011, http://advancinghumanrights.org/news/police_in_iran_destroy_satellite_dishes. Last accessed 2011–09–01.
11. 'Satellite Dishes as Decorative Objects', posted by Cory Doctorow, *Boing Boing*, June 26, 2007. Available at http://www.boingboing.net/2007/06/26/satellite_dishes_as_.html. Last accessed 2008–04–15.
12. 'Satellite Dish Art', *Dark Roasted Blend: Weird and Wonderful Things*, July 13, 2007. Available at www.darkroastedblend.com/2007/07/fun-with-satellite-dishes.html. Last accessed 2010–05–15.

REFERENCES

Biemann, U. 2005. Black Sea Files. In *B-Zone: Becoming Europe and Beyond*, edited by A. Franke. Barcelona: Actar Press, 16–95.

Bostom, A. 2008. Geert Wilders: Wisdom and Courage. *EuropeNews*, September 29. Available at http://europenews.dk/en/node/14505. Last accessed 2010–05–01.

Briel, R. 2011. Satellite Reception Confirmed to be a Human Right. *BroadbandTV News*, August 19. Available at http://www.broadbandtvnews.

com/2011/08/19/satellite-reception-confirmed-to-be-a-human-right/. Last accessed 2011–09–02.

Brunsdon, C. 1997. Satellite Dishes and Landscapes of Taste. In *Screen Tastes*. London: Routledge, 148–164.

Buruma, I. 2005. Letter from Amsterdam: Final Cut. *The New Yorker*, January 3. Available at http://www.newyorker.com/archive/2005/01/03/050103fa_fact1. Last accessed 2010–05–26.

Carey, J. 1992. *Communication as Culture*. London: Routledge.

Drury, I. 2011. Having a Satellite Dish is a 'Human Right', says European Court. *Daily Mail Online* (UK), August 8. Available at http://www.dailymail.co.uk/news/article-2023359/Having-satellite-dish-human-right-says-European-court.html. Last accessed 2011–09–02.

Fiske, J. 1996. *Media Matters*. Minneapolis: University of Minnesota Press.

Graham, S. and S. Marvin. 2001. *Splintering Urbanism: Networked Infrastructures, Technological Mobilities and the Urban Condition*. London: Routledge.

Hargreaves, A. and D. Mahdjoub. 1997. Satellite Television Viewing among Ethnic Minorities in France, *European Journal of Communication*, 12(4), 459–477.

Hayes, B. 2005. *Infrastructure: A Field Guide to the Industrial Landscape*. New York: W. W. Norton & Company.

Kilner, J. 2011. Turkmenistan Detains Bloggers Who Reported on Arms Blast. *The Telegraph*, August 12. Available at http://www.telegraph.co.uk/news/worldnews/asia/turkmenistan/8697121/Turkmenistan-detains-bloggers-who-reported-on-arms-cache-blast.html. Last accessed 2011–09–01.

Latour, B. and P. Weibel. 2005. From Real Politik to Dingpolitik or How to Make Things Public. In *Making Things Public: Atmospheres of Democracy*, edited by B. Latour and P. Weibel. Cambridge, MA: MIT Press, 14–43.

Leiken, R. S. 2005. Europe's Angry Muslims. *Foreign Affairs*, July–August. Available at http://www.cfr.org/publication/8218/. Last accessed 2010–05–26.

Morley, D. 2000. *Home Territories: Media, Mobility, and Identity*. London: Routledge.

Parks, L. 2007. Where the Cable Ends: Television in Fringe Areas. In *Cable Visions: Television Beyond Broadcasting*, edited by S. Banet-Weiser, C. Chris and A. Freitas. New York: New York University Press, 103–126.

Parks, L. 2009. Around the Antenna Tree: The Politics of Infrastructural Visibility. Flow, March 2009. Available at http://flowtv.org/?p=2507. Last retrieved 2011–09–27.

Price, M. 2002. Satellite Broadcasting as Trade Routes in the Sky. In Search of Boundaries: Communication, Nation States and Cultural Identities, edited by J. M. Chan and B. T. McIntyre. Westport, CT: Ablex Publishing: 146–166.

Pridemore, A. 2005. Dishing It Out: Man Beams 5000 Radio, TV Channels with a Dozen Satellite Receivers. *The Register-Herald* (Beckley, WV), November 27. Available at http://www.register-herald.com/homepage/x519054803/Dishing-it-out. Last accessed 2010–05–15.

Saidazimova, G. 2007. Turkmenistan: A New Obstacle for Access to the Airwaves. *Radio Free Europe, Radio Liberty*, December 5. Available at http://www.rferl.org/content/article/1079232.html. Last accessed 2011–09–01.

Sakr, N. 2002. *Satellite Realms: Satellite Television, Globalization and the Middle East*. London: I.B. Tauris.

Sang-Hun, C. 2009. Rural South Korean's Global Links Grow, Nourished by a Satellite Crop. *The New York Times*, June 29. Available at http://www.nytimes.com/2009/06/30/world/asia/30satellite.html. Last accessed 2010–05–15.

Saterangsee, S. 2010. Satellite Dishes made by Students Sell like Hotcakes. *The Nation* (Thailand), January 17. Available at www.nationmultimedia.com/

Part II

Epistemologies of Cultural Technologies

6 Being 'Accountable'

TV Audiences and Surveillance

Toby Miller

> In most societies in which supernatural elements are important in attaining success, some form of divination is practiced, because foreknowledge is one way of control. In parts of East Africa, the entrails of chickens are used for divining the future, while among the Karen of Burma it is the gall bladder of a pig; in Hollywood polls are used to determine the mysterious tastes of the audience. (Powdermaker 1950: 285)

> We all know the power of television to make us laugh, cry, hide behind the sofa or forget our worries, particularly as we spend more time at home trying to escape the doom and gloom of the credit crunch. To help the nation seek some solace in February, one of the most depressing months, psychologist Donna Dawson has put together her pick of the month's programmes on Sky+HD to help lift your spirits. (Dawson 2009)

> TV is still the 800 pound gorilla of media consumption, with viewers glued to their sets over 158 hours per month. In fact, TV viewing has increased more than an hour a day in the last decade, far outpacing any other major media option. ('Data Integration' 2010)

The first epigraph quotes the noted anthropologist who gave the *soubriquet* the 'Dream Factory' to Hollywood after a year's painstaking ethnography in the wilds of the studios. Powdermaker was referring to the industry's desperate desire to know and control audiences.

The second epigraph comes from a psychologist employed by Britain's leading satellite provider of television, Sky. It originally appeared as part of the company's 'Happier Days TV Guide'. Dawson's pop-therapeutic alibi was a means of corralling viewers into particular categories,[1] the better to know them and urge them to buy high-definition sets, which were thought to 'improve the psychological well being of the viewer' ('Sky+HD' 2009).

The third epigraph is the work of a noted TV ratings firm: 'In 100 countries around the world, Nielsen provides clients the most complete understanding of what consumers watch and buy'.[2] Nielsen seeks to counter the popular mythology of the *bourgeois* media (and those quaint academic narcissographers who love to write about themselves or their families) that television no longer matters.

Together, these quotations encapsulate the core message of this chapter: Whether we see audiences as knowledgeable, ignorant, passive, active, powerful, weak, intelligent, idiotic—or none, all or some of the above—their principal role for the television industry is to be known, and that role remains vital across the media landscape. As such, their active participation in watching texts or creating paratexts is valuable to the industry, because it offers corporations greater surveillance of them. For it is audiences' attention, their mental and emotional labour, that gives them value. The more visible that labour becomes—the more that is known about the audience's composition and conduct—the better off corporate capitalism is. I will demonstrate this with a case study from US TV.[3]

INTRODUCTION

Audiences are not already-extant entities participating in the neoliberal wet dream of supply and demand. Most of the time, television is not directly selling to its audience, and the audience is not buying from it. Fifty years ago, Dallas Smythe (2004: 319f) explained that audience attention was the key commodity that television stations sold to advertisers, even though it was not material and was difficult to define and measure. TV texts are not so much commodities as 'symbols for time' (Hartley 1987: 133). In fact, the industry positions itself 'at the intersection of time and real estate' (Maggio 2008). To do so, it routinely undertakes surveillance (Gandy 1989).

Despite Nielsen's best efforts, any discussion of TV nowadays has to deal with the argument that the medium has had its day. The rhetoric surrounding newer audiovisual media is inflected with the phenomenological awe of a precocious child set to heal the wounds of modern life, magically reconciling public and private, labour and leisure, commerce and culture, citizenship and consumption. The alleged upshot? *La fin de la télévision* (The end of television) (Missika 2006); *La televisión ha muerto* (Television is dead) (De Silva 2000). The grand organizer of daily life over half a century has supposedly lost its pride of place in both the physical layout of the home and the daily order of drama and data. We must all say 'Bienvenidos al mundo de la postelevisión' (Welcome to the post-television world). Dual monopolies have allegedly been broken: The physical object no longer dominates, nor does its model of unidirectional production (Verón 2008).

Historically, most new media have supplanted earlier ones as central organs of authority or pleasure, as per books versus speeches, films versus plays and records versus performances. The present conjuncture is meant to be one more proof of that teleology. The corporate recorded music industry is shrinking due to online exchange, with double-digit decreases in compact disc sales over the last decade. The Internet is displacing newspapers in the Global North as a key source of political and commercial information (Friedman 2008; Pew Research Center 2008).

But TV is unusual. It blended earlier media to become a warehouse of contemporary culture (Newcomb 2005: 110; Standage 2006). Television modelled the Internet and commerce conceptually, just as it modelled the personal computer stylistically. And the early signs are that the Internet and TV transform one another. In the words of Steve Ballmer, Microsoft's chief executive, '[W]e will see TVs become more sophisticated and more connected. The boundary between the PC and the TV will dissolve' (quoted in Moses 2009; also see Microsoft 2009).

Even cybertarian true believers, who patronisingly refer to television as part of 'legacy media', must confront the fact that the sites most people visit are really rather distant from futurist fantasies. In 2010, the US broadcast networks' websites increased their viewership by an average of 82%, and twenty-three of the top twenty-five Internet news destinations were professional outlets from before the web began. Worldwide, the power of the Internet to attract audiences is clearest in the deployment of venerable sources and genres: the BBC (employing a lot of properly trained journalists as per the 20th century), YouTube (featuring material 'liberated' from film and TV as per the 20th century) and Wikipedia (following an 18th century format) ('A Cyber-House' 2010; Kelly 2008; comScore 2011; Rosenstiel 2011; Sacerdoti 2011).

And US TV continues to grow by every indicator imaginable. In 2006, more than 98% of US homes had at least one set, while 64% subscribed to cable, up twenty points in twenty years. Consumers spent $20 billion buying new TVs that year. By 2007, 51% of people owned three or more, and the next year, the number of US households owning televisions increased by 1.5%, with particularly significant growth among migrants and their recent descendants (Spangler 2009; Motion Picture Association of America 2007: 35 and 37; Borland and Hansen 2007; Ellis 2007: 40; Reynolds 2009).

The average US resident devoted 127.25 hours to television a month in 2006, as opposed to 26.5 hours online and 2.25 hours with their cell phones; in 2009, 93% of adults watched at least an hour of TV a day, but just 4% consumed an hour of video online each day. Three quarters of people did so at some time—but they spent seventy times more hours a month doing so via a conventional set ('Nielsen Media' 2006; 'Majority of Americans' 2007; Shields 2009).

And the young? People born between 1984 and 1990 choose television over the Internet and the cell, for both entertainment and information. Half the Internet sites that children aged between 6 and 11 visit attract their attention through advertising on TV or in print. US children between the ages of 6 and 14 tune to television at rates unprecedented for twenty years; 69% have sets in their bedrooms, versus 18% with Internet access and 49% owning or subscribing to videogames. Those aged between 2 and 11 watched 17.3[4] hours of TV a week in 2006, an increase on the previous year. The keenest viewers are young girls. They quite like new technology, and adopt it at a frenetic pace—but 'TV is king', in the words of The Tubes' album *Remote Control* (1979) (Pew Research Center 2005; Downey 2007; Friedman 2009b).

Across the United States, television is the most influential advertising medium, and its influence is greater than during the pre-web period. Hundreds of case studies undertaken over the past two decades confirm that TV is the principal source of raising brand awareness. Of the top hundred companies recognized by residents in 2009, sixteen were television or film related, with CNN and MTV in the top ten (Loechner 2009b; Neff 2009; Thomasch 2009; Reynolds 2009).

In short, as the web matures, more and more Gringos watch TV (Umstead 2010). Television still occupies vast amounts of people's time and money, because it delivers information and entertainment with astonishing speed and ease. And its audience is subject to increasing fascination and knowledge, a creature for inquiry through surveillance.

ACCOUNTING AND SURVEILLANCE

The US television industry would not utilise my chosen vocabulary of surveillance to describe its activities. Television mavens are much more enamoured by 'accountability'. That concept seems a long way from the political issues raised by surveillance, and equally distant from the fun that is deemed central to the medium. It is worthy of critical examination.

The far-from-thrilling nature of accountancy has been established ethnographically, lexicographically and rhetorically (Flowerdew and Wan 2006), as per Monty Python's renowned vocational-guidance 'Lion Tamer' sketch ('dull, dull, dull') featuring the League for Fighting Chartered Accountancy. Desperate to reverse, or perhaps to play on, these longstanding connotations, the Californian accounting firm MGO adopted 'Proud to Be Boring Accountants' as its slogan in 2010.4 You do not marry such people for the ride of your life. And in this case, their work is being used as a euphemism.

Accountability *should* refer to corporations and governments being accountable to popular democracy and financial auditing, with their representational norms, profit-and-loss columns and hiring practices publicly available for scrutiny. But in TV, the term signifies the amount of information about audiences that networks hand to advertisers—what kinds of people watched and what they went on to purchase. Accountability equals surveillance.

Surveillance has long been a central strut of modernity, supposedly to make populations secure, content and productive. Foucault explains the phenomenon like this:

> [A]n important problem for [French] towns in the eighteenth century was allowing for surveillance, since the suppression of city walls made necessary by economic development meant that one could no longer close towns in the evening or closely supervise daily comings and goings, so that the insecurity of the towns was increased by the influx of the

> floating population of beggars, vagrants, delinquents, criminals, thieves, murderers, and so on, who might come, as everyone knows, from the country . . . In other words, it was a matter of organizing circulation, eliminating its dangerous elements, making a division between good and bad circulation, and maximizing the good circulation by diminishing the bad (Foucault 2007: 54; also see Foucault 1976; Ravel 1999)

With the expansion of state authority into the everyday, into all corners of life, the *quid pro quo* for the security afforded by governments became knowing everyone's identities and practices. The equivalent expansion of corporations into the everyday, into the same corners of life, had as *its quid pro quo* for the provision of goods and services that they, too, know more and more about us. The proliferation of methods for studying how people interact with the media may even infringe privacy and free speech, given the proprietorial methods used by corporations to undermine autonomous reading (Turow 2005; Andrejevic 2006; Baruh 2004). The World Privacy Forum proposes that we are in a 'one-way mirror society', where power accretes to corporations through the supposedly even-handed tool of interactivity (Dixon 2010). It is certainly a good time for commodifying and governing talents, interests and ideas by transforming them into intellectual property.

US TV is available to spectators all day, every day. The price is more than purchasing consumer electronics, subscribing to satellite or cable, or paying for the electricity to drive these appliances. It is the price of having your identity known, governed, bought, and sold:

> What do Klingons, giant rubber bands, and monster trucks all have in common? They hold the fascination of every male on the planet—and they make frequent appearances on Spike TV. Spike TV, which totes itself as a network for men, is the perfect example of the power of knowing your audience and giving them what they want.
>
> In one week of watching Spike TV you can expect to see poker, ultimate fighting, demolition derbies, cop shows, kung fu movies, school bus figure eight racing, pyrotechnic explosions, Klingons, customized cars, Chairman Tamori, giant rubber bands, sports superstars, Captain James T. Kirk, video game awards, Japanese trivia shows, dating competitions, MacGyver, and monster trucks. And that is only a partial list of what Spike TV offers men. (Smith n.d.)

AUDIENCE FANTASIES, AUDIENCE NUMBERS

Three basic systems of fantasizing about TV audiences dominate marketing: the individual, the regional and the global. The first is animated by classifications of race, class, gender, age and psyche; the second by geopolitical

clusters; and the third by a growing cosmopolitanism. Audience surveillance starts before television programs have even been made, via focus groups, which scrutinise small numbers of people whose identities represent the social formations desired by advertisers. They are shown pilots of programs to judge the likelihood that people like them will watch shows that are picked up by networks. These groups are crucial to the arrival and departure of US TV texts. The firms that undertake such research frequently work for both producers and networks, thereby creating a conflict of interest, and are rarely multilingual, unlike their equivalents in public relations or advertising.

Once a series is on the air, ratings and subscriptions are the keys to determining their success. They have become of increasing importance as the audience for television has fractured. The English-language US broadcast networks attained their peak viewing numbers in 1976, with 92% of the national audience; by 2005, they had 45% of it.[5] US cable programming thrived at their expense with the deregulation of the 1980s and 1990s, which permitted cable networks to start their own stations and make their own programs. Today's highly targeted cable networks offer original programming. They seek a signature in the public mind and correlate consumption with viewing. The upshot? Perhaps most spectacularly, more people watched CNN than any broadcast network on election night in 2008, and no fewer than thirty-seven cable stations that carry commercials reported their best prime-time viewing figures that year, while audience numbers for the broadcast networks dropped by an average of 11% (Attallah 2007: 330; Flaherty 2008; Richardson and Figueroa 2005; Collins 2009; Hassan et al. 2003: 446f; Morris 2007).

Companies like MTV attract young audiences who are less inclined to sit through commercials than their elders. Some research suggests they customise TV schedules to suit their diaries, not to avoid advertising. So now there are more and more ideas about interactive commercials, where viewers use remote controls to respond to pitches—and disclose data about themselves (Helm 2007; Downey 2007; Reynolds 2008; Edgecliffe-Johnson 2007).

Some telling examples of intense audience targeting and surveillance come from sporting TV. At the height of the Great Recession, the premium-cable network HBO had 41 million US subscribers paying $3.84 billion a year. Their social identities and viewing interests are very precisely calibrated (Ben Block 2010). For example, visitors to HBO's web site on boxing encounter a section titled 'TALK', which invites them to participate in polls, sign up for a newsletter and write on bulletin boards.[6] 'TALK' is also and equally a system of surveillance that allows the network to monitor viewers' tastes and ideas without paying them for their intellectual property (Miller and Kim 2008). The basic-cable sports network ESPN uses interactive TV fora such as 'My Vote' and 'My Bottom Line' to uncover audience drives in the name of enabling participation and pleasure in watching. ESPN has also purchased broadband portals that ensure global dominance.

It now owns Cricinfo, Scrum.com and Racing-Live, which provide two-way exchanges with their audiences to build loyalty and deliver intelligence to advertisers (Nordyke 2008; Hampp 2008; Spangler 2009; Gibson 2009; Stiernstedt 2008).

For its part, Fox Soccer Channel succeeds not because it commands huge audiences, but because of their composition—men aged 18–34 with annual household incomes of more than $75,000 ('Fox Soccer' 2008). The station's biggest-ever ratings, for a 2011 English Premier League game, drew 418,000 average viewers and 285,000 households.[7] That may not seem significant in a country of over 300 million people, but with a highly targeted demographic strategy, such numbers can produce profitable outcomes. The hint as to how the channel identifies viewers to advertisers comes from its TV commercials, which are about regaining and sustaining hair growth and hard-ons, losing and hiding pimples and pounds, and becoming and adoring soldiers and sailors.

And broadcast TV? A network drama like *Alias* (2001–2006), which did not rate well, remained on the air for five years because of the youthfulness of its fans and its successful promotion of DVD sales. Low-rating situation comedies about élites, like *30 Rock* (2006–) and *The Office* (2005–), keep their places on the schedule as a consequence of being much loved by affluent viewers. That quality can enable unpopular series to survive, because advertisers of costly merchandise are promised ruling-class audiences by networks. With overall declines in ratings, the broadcast networks are sometimes forced to offer free commercial time to advertisers who have paid for programs that do not attract the 'right' people. Then there is very specific, local targeting, where broadcast stations follow the lead of radio and the airlines, via credit and debit cards articulated to frequent viewing and rewards systems with local advertisers (Downey 2007; Consoli 2008; Greenwald 2009).

Youthful audiences are of particular surveillant concern to the industry, because the young are thought to be still deciding on their favourite commodities—toothpaste, transport, tutus, and so on. To quote 1970s' ABC executive Leonard Goldstein, they are 'the most curious' viewers, the likeliest to 'seek out the new'. The success of *Friends* (1994–2004) encouraged broadcast networks to schedule comedies that would appeal to people in their twenties and thirties. Other age groups got the message that they were not a priority and did the sensible thing of turning to cable stations. Ironically, when CBS reverted to the idea of addressing a mass audience in 2008, it won the ratings both overall *and* among young people (Collins 2009).

Broadcasters are forever announcing new, failsafe schemes for captivating and capturing the audience. In 2006, NBC unveiled 'Television 2.0', which was meant to be the end of drama in prime time. In 2008, it declared the return of the '8 o'clock Family Hour' with serial drama throughout the year—this was called 'The New Paradigm'. Then 2009 ushered in 'The NBCUniversal2.0', a 'New, New Paradigm' with less original programming and more reality and talk shows, described in the idiotic vocabulary of managerialism as a 'margin enhancer'. Consider this embarrassing quotation from the head of NBC-Universal's TV and movie interests in 2009: 'We have a sniper focus on 8 p.m. to 10 p.m. to drive a power audience flow' (de Moraes 2008; Friedman 2009a). The translation is that the network had given up on creating high-quality drama other than during those two hours.

It is hard to believe that these people inhabit the world of English when they use such laughably inelegant and aggressive metaphors to describe the surveillance and management of viewers. Then the industry turns the manifold, manifest failures of these manager-warlocks into assertions that audiences are their masters!

I do not suggest that the practice of surveillance is omniscient and omnipotent. Audience numbers and practices have massive effects, but they are not pure, unvarnished accounts of popularity and conduct. For example, ratings apartheid was practised for decades in the US until 2007, initially because Spanish-language networks thought their viewers were being lost in the Anglo-Yanqui mass. When Spanish-language networks were finally measured alongside their English-language equivalents, Univision won the ratings amongst advertising's most desired age group—18–49—no fewer than fourteen times in 2008, because Latin@s were not departing network TV for cable or the Internet, due to their economic situation (Bauder 2008).[8] In 2012, the network provided English-language subtitles to its programs, thereby locking in multi-generational families with varying linguistic competences.

The belated recognition of Univision's importance emphasises the limitations of ratings. The measurement of bilingual audiences to Anglo networks was so incompetent that it was wrongly used to downplay the appeal of ethnically inclusive English-language material, misreading viewers' desires and hence diminishing work prospects for minority talent. Numerous multicultural shows were prematurely cancelled, such as *Greetings from Tucson* (2002–2003), *Kingpin* (2003) and *Luis* (2003), because their audiences were underestimated—in every sense. What else were Spanish speakers to do but turn to Univision, when Anglo networks systematically ignored, distorted or misunderstood them, as decades of content analysis in the *National Brownout Report* have illustrated (National Association 2006; Rincón & Associates 2004; James 2007)? This connects to the intellectual narrowness of Anglo executives and attempts by traditional networks to minimise the power of cable stations by stressing prime time as the centrepiece of measuring audience size, which suited their heavy investments in marquee programming at that time, and their cosy relationship with the companies that research viewers.

The panic over Univision's success and the gains made by cable stations over broadcast networks repeat the anxiety of the 1980s, when video-cassette recorders and premium channels threatened to take control of the audience away from TV corporations and towards film studios, cable companies and viewers. These concerns were reawakened when Digital Video

Recorders (DVRs) permitted viewers to elude the clutches of capital by avoiding commercials in real time. DVRs were even advertised for these qualities, supposedly transforming audiences into schedulers. There was a dramatic loss of confidence in TV's efficacy amongst major advertisers, despite this additional intelligence about viewers, out of fear that '[l]e pouvoir de programmer passe des mains de l'éditeur à celles du téléspectateur' (the power to program is shifting from editors to spectators) (Cristiani and Missika 2007; also see Carlson 2006).

By 2009, almost a third of US homes had DVRs. But just 5% of TV was being time shifted, and people skipped a mere 3% of commercials ('The Revolution' 2009). Time shifting grew in 2010, with the average resident watching nearly ten and a half hours more that way by the end of the year. The biggest annual increase was in the third quarter, which was 17.9% over the same period in 2009.[9] But these increases came from a small base. Furthermore, the first popular versions of these devices only worked when subscribers connected them to the Internet, thereby allowing service providers, TiVo and ReplayTV, to collect information. In addition to amassing a huge database of consumer information, they pinpointed the identities and actions of television viewers (Lewis 2001: 40; Rose 2001; Attallah 2007: 330; 'The Revolution' 2009).

To repeat, the elemental desire that drives advertisers is not absolute numbers of viewers. They want information about, and surveillance of, those audiences in terms of identity, wealth and taste. Ratings firms have developed ever-more impressive-sounding methods of investigating audiences—10,000 US viewers are under surveillance through People Meters nowadays, to add to the 15,000 examined by other means, such as Anytime Anywhere Media Measurement. And online sites replaying network television and movies, such as Hulu, use 'geo-filtered access logs' to identify audiences. These are measured each day, alongside confessional testimonies by potential viewers—if you tell us about your life and practices of consumption, we'll tell you about programs that may interest you ('Nielsen Media' 2006; Mermigas 2008). This is the future of 'TV Everywhere' initiatives.

Consider YouTube's Video Identification software, which was developed with Disney and Time Warner. It is a surveillance device for tracking copyrighted materials on the site that follows the history of each uploaded frame, spying on users to disclose their Internet protocols, aliases and practices to corporations. The software permits these companies to block or enable reuse of texts, depending on their marketing and surveillance needs of the moment. YouTube has thus become commercial TV's valued ally, tracking intellectual property and realising the culture industries' paradoxical dream of engaging in product placement each time copyright is infringed on line, while learning more and more about their audiences (Miller 2009).

This passion for knowing the customer at the same time as claiming to serve it explains the advent of firms such as Phorm and FrontPorch ('Watching' 2008). Corporate consultant Openwave's *Privacy Primer* says it protects consumers from an *Era of Behavioral Marketing*. But the *Primer* gleefully avows that '[o]n the internet, customer feedback isn't requested so much as it's collected, like a digital trail of breadcrumbs. Mobile technology only sharpens the focus on user behavior by bringing location and contextual information into play' (Openwave 2009).

CONCLUSION

Focusing on corporate power, as I have done here, leaves one open to the venerable, predictable charge of denying the ability of audiences to function autonomously of how the TV industry wants them to behave. Foucault rightly said, 'On se plaint toujours que les médias bourrent la tête des gens. Il y a de la misanthropie dans cette idée. Je crois au contraire que les gens réagissent; plus on veut les convaincre, plus ils s'interrogent' (Some complain that the media brainwash people. This seems misanthropic to me. I believe that people resist; the more one tries to convince them, the more they question) (Foucault 2001: 927). Stuart Hall concurs: 'I speak and talk to the radio and the TV all the time. I say, "that is not true" and "you are lying through your teeth" and "that cannot be so". I keep up a running dialogue' (Taylor 2006).

Very nice. But lest it be assumed that ratings and so on are purely the stuff of old-style effects studies, we should recognise that marketers like nothing better than active audiences who are overflowing with knowledge about programs; nothing better than diverse, articulate groups with easily-identified cultural politics and practices; nothing better than fine-grained ethnographic and focus-group work in addition to large-scale surveys that provide broad-based demographic data. The supposedly resistive social identity is just one more category for their delectation (Maxwell 1996a, 1996b).

And training students to analyse TV texts is in no way threatening to commerce: 'Advertisers think media education is great' (Bazalgette 1999). No wonder Virginia Postrel's (1999) infamous *Wall Street Journal* op-ed welcomes media and cultural studies as 'deeply threatening to traditional leftist views of commerce . . . lending support to the corporate enemy and even training graduate students who wind up doing market research'. Beyond our own field, capitalist lackeys in neuroscience lie in wait to scan audience brains and identify which segments activate purchasing desires while viewing (Bürgi 2007).

Marketers avow their powerlessness over audiences when challenged in the public sphere, but boast omnipotence over them in the private world: the essay that won the oleaginous 'Best New Thinking Award' at the 2003 Market Research Society Conference acknowledged that successful marketing does not 'view . . . the consumer as an individual' but 'part of the herd' (Earls 2003). And producers who hide behind the rhetoric of sovereign consumption frequently hold audiences in contempt. The creator of

Heroes (2006–2010) referred to people who viewed his show on network TV as 'saps and dipshits who can't figure out how to watch it in a superior way' (quoted in Hirschorn 2009). Thanks for that.

The task for those of us who want television to be more entertaining, informative, inclusive and democratic is to alert our fellow audience members that our viewing and social identities are governed and commodified and our mental labour sold to others without our consent or remuneration. We must make TV accountable by exposing its own surveillant eye to scrutiny by those it seeks to observe and channel.

NOTES

1. http://www.donnadawson.co.uk. Last accessed 2011–10–07.
2. http://www.nielsen.com/us/en/about-us.html. Last accessed 2011–10–07.
3. Thanks to Rick Maxwell for first alerting me to thinking in the way that animates this chapter and to the editor for his helpful and supportive comments and his understanding of how to manage my peccadillos!
4. http://www.mgocpa.com/go/mgo/;jsessionid=EC11314641BFF963CA5F3F1 4BBCDE75F. Last accessed 2011–10–07.
5. CBS, ABC, NBC in 1976; Fox appeared later.
6. http://www.hbo.com/boxing/index.html#/boxing/talk/index.html. Last accessed 2011–10–07.
7. http://www.epltalk.com/fox-soccer-channel-sets-new-record-for-chelsea-vs-liverpool-telecast-29232. Last accessed 2011–10–07.
8. This prompted one more tedious but nasty turn in a national debate over assimilation that saw ludicrous accusations to the effect that young Latin@s were not learning English or patriotic identification as a consequence of watching shows in another tongue (Arnoldy 2007). Empirical studies of minority groups using TV to solidify their culture and remain in touch with places of origin, such as Turks in Greece or Arabs in the US, counter the notion that this precludes integration, and Latin@s move easily between languages, code-switching both inter-sententially and between phrases, in keeping with their use of both Anglo and Spanish television channels (Madianou 2005: 55; Rizkallah and Razzouk 2006; Loechner 2009a).
9. http://blog.nielsen.com/nielsenwire/media_entertainment/tv-usage-trends-q3-and-q4–2010. Last accessed 2011–10–07.

REFERENCES

Andrejevic, M. 2006. Total Information Awareness: The Media Version. *Flow* 4(8). Retrieved from http://flowtv.org/2006/07/total-information-awareness-the-media-version. Last accessed 2011–10–07.

Arnoldy, B. 2007, September 17. Among Networks, Spanish-Language Univision is now a Top Contender. *Christian Science Monitor.* Retrieved from http://www.csmonitor.com/2007/0917/p01s03-ussc.html. Last accessed 2011–10–07.

Attallah, P. 2007. A Usable History for the Study of Television. *Canadian Review of American Studies/Revue canadienne d'études américaines*, 37(3), 325–349.

Baruh, L. 2004. Anonymous Reading in Interactive Media. *Knowledge, Technology, & Policy* 17(1), 59–73.

Bauder, D. 2008. Univision Takes 18–49 Demo Crown. *TVNewsday*, December 31. Retrieved from http://www.tvnewscheck.com/article/2008/12/31/28301/univision-takes-1849-demo-crown. Last accessed 2011–10–07.

Bazalgette, C. 1999. Sabre-Toothed Tigers and Polar Bears. *Media Education Journal.*

Ben Block, A. 2010. Pay TV Paying Off in Down Economy. *Hollywood Reporter*, July 1. Retrieved from http://www.reuters.com/article/2010/07/02/industry-us-paytv-idUSTRE66110E20100702. Last accessed 2011–10–07.

Borland, J. and E. Hansen. 2007. The TV is Dead. Long Live the TV. *Wired*, April 6. Retrieved from http://www.wired.com/entertainment/hollywood/news/2007/04/tvhistory_0406. Last accessed 2011–10–07.

Bürgi, M. 2007. Leveraging Brain Power. *MediaWeek*, September 24, 9.

Cabletelevision Advertising Bureau. 2007. *Power of the Young*. New York: CAB.

Carlson, M. 2006. Tapping into TiVo: Digital Video Recorders and the Transition from Schedules to Surveillance in Television. *New Media & Society*, 8(1), 97–115.

Collins, S. 2009. TV Starts to Notice Others in the Room. *Los Angeles Times*, January 11. Retrieved from http://articles.latimes.com/2009/jan/11/entertainment/et-demographic11. Last accessed 2011–10–07.

comScore. 2011. *comScore 2010 U.S. Digital Year in Review*, February. Retrieved from http://www.comscore.com/Press_Events/Presentations_Whitepapers/2011/2010_US_Digital_Year_in_Review. Last accessed 2011–10–07.

Consoli, J. 2008. TV Retains Marketing Dollars in Hard Times. *New York Times*, December 30. Retrieved from http://www.nytimes.com/2008/12/30/business/media/30adco.html. Last accessed 2011–10–07.

Cristiani, M. and J.-L. Missika. 2007. Internet ne fait pas disparaître le lien social, il donne aux individus la possibilité de le maîtriser davantage. *L'Atelier*, April 12. Retrieved from http://www.atelier.net/fr/node/180667/summary/227764. Last accessed 2011–10–07.

A Cyber-House Divided. 2010. *Economist*, September 4, 61–62.

Data Integration Optimizes Advertising Media Mix. 2010. *Nielsen Wire*, July 1. Retrieved from http://blog.nielsen.com/nielsenwire/online_mobile/data-integration-optimizes-advertising-media-mix. Last accessed 2011–10–07.

Dawson, D. 2009. Psychologist's Choice. *SKY TV*. Retrieved from http://hdbt.com/a489399-sky-hd-reveals-high-definition-programming-can.cfm. Last accessed 2011–10–07.

de Moraes, L. 2008. At NBC, There is no Script for Success. *Washington Post*, December 30, C1.

De Silva, J. P. 2000. *La televisión ha muerto: La nueva producción audiovisual en la era de Internet: La tercera revolución industrial*. Barcelona: Editorial Gedisa.

Dixon, P. 2010. *The One-Way Mirror Society: Privacy Implications of the New Digital Signage Networks*. World Privacy Forum. Retrieved from http://www.worldprivacyforum.org/pdf/onewaymirrorsocietyfs.pdf. Last accessed 2011–10–07.

Downey, K. 2007. In Their TV Tastes, the Rich are Different. *Media Life*, March 5. Retrieved from http://www.medialifemagazine.com/artman/publish/article_10573.asp. Last accessed 2011–10–07.

Earls, M. 2003. Advertising to the Herd: How Understanding Our True Nature Challenges the Ways We Think about Advertising and Market Research. *International Journal of Market Research*, 45(3), 311–337.

Edgecliffe-Johnson, A. 2007. Targeted Television Ads Three Years Away. *Financial Times*, October 30. Retrieved from http://www.ft.com/cms/s/0/587c32c2–868b-11dc-b00e-0000779fd2ac.html#axzz1GLfDgcZl. Last accessed 2011–10–07.

Ellis, J. 2007. *TV FAQ*. London: IB Tauris.

Flaherty, M. 2008. Turner Unveils Television Statistics. *Variety*, December 10. Retrieved from http://www.variety.com/article/VR1117997186?refCatId=18. Last accessed 2011–10–07.

Flowerdew, J. and A. Wan. 2006. Genre Analysis of Tax Computation Letters: How and Why Accountants Write the Way They Do. *English for Specific Purposes*, 25(2), 133–153.

nd1,Littlista>Foucault, M. 1976. *Histoire de la sexualité, I: La Volonté de savoir*. Paris: Gallimard.

Foucault, M. 2001. *Dits et écrits 1954–1988. II: 1967–1988*. Edited by Daniel Defert and François Ewald with Jacques Lagrange. Paris: Quarto Gallimard.

Foucault, M. 2007. Spaces of Security: The Example of the Town, Lecture of 11th January 1978. Translated by Graham Burchell. *Political Geography*, 26(1), 48–56.

Fox Soccer Channel Becomes an Official Nielsen[-]Rated Network Starting October 1. 2008. *Foxsports.com*, September 15. Retrieved from http://www.pr-inside.com/fox-soccer-channel-becomes-an-official-r807097.htm. Last accessed 2011–10–07.

Friedman, W. 2008. Digital Lessons TV Should Learn from Music Companies: Watch the Rising Waters. *MediaPost's TV Watch*, November 26. Retrieved from http://www.mediapost.com/publications/?fa=Articles.showArticle&art_aid=95528. Last accessed 2011–10–07.

Friedman, W. 2009a. NATPE Executives Call for Risk-Taking with 'Laser Focus'. *MediaPost's TV Watch*, January 28. Retrieved from http://www.mediapost.com/publications/?fa=Articles.showArticle&art_aid=99270. Last accessed 2011–10–07.

Friedman, W. 2009b, April 21. Study: Teens Love Live TV. *MediaDailyNews*. Retrieved from http://www.mediapost.com/publications/?fa=Articles.showArticle&art_aid=104573. Last accessed 2011–10–07.

Gandy, O. H., Jr. 1989. The Surveillance Society: Information Technology and Bureaucratic Control. *Journal of Communication*, 39(3), 61–76.

Gibson, O. 2009. Where Will the Big Bucks Go When the Bubble Bursts? *Guardian*, January 22. Retrieved from http://www.guardian.co.uk/sport/2009/jan/22/owen-gibson-sport-on-tv-economic-crisis. Last accessed 2011–10–07.

Greenwald, A. 2009. KVVU Rewards Build Ratings, Revenue. *TVNewscheck*, March 2. Retrieved from http://www.tvnewscheck.com/article/2009/03/02/30005/kvvu-rewards-build-ratings-revenue. Last accessed 2011–10–07.

Grindstaff, L. and J. Turow. 2006. Video Cultures: Television Sociology in the 'New TV' Age. *Annual Review of Sociology*, 32, 103–25.

Hampp, A. 2008. ESPN Rolls Out Interactive Features. *Advertising Age*, December 23. Retrieved from http://adage.com/print?article_id=133463. Last accessed 2011–10–07.

Hartley, J. 1987. Invisible Fictions: Television Audiences, Paedocracy, Pleasure. *Textual Practice* 1(2), 121–138.

Hassan, S.S., S. Craft and W. Kortam. 2003. Understanding the New Bases for Global Market Segmentation. *Journal of Consumer Marketing* 20(5), 446–462.

Helm, B. 2007. Cable Takes a Rating Hit. *Business Week*, September 24.

Hirschorn, M. 2009. The Future is Cheese. *Atlantic Monthly*, March. Retrieved from http://www.theatlantic.com/magazine/archive/2009/03/the-future-is-cheese/7277. Last accessed 2011–10–07.

James, M. 2007. Nielsen Ends Separate Latino TV Survey. *Los Angeles Times*, August 27, C1, C4.

Kelly, J. 2008. *Pride of Place: Mainstream Media and the Networked Public Sphere*. Cambridge, MA: Berkman Center for Internet and Society at Harvard University.

Lewis, G. 2001. Think Local When Going Global. *Variety*, February 26–March 4, 7.

Loechner, J. 2009a. Bi-lingual Hispanics Live with Ease in Both Worlds. *MediaPost*, February 16. Retrieved from http://www.mediapost.com/publications/?fa=Articles.showArticle&art_aid=100359. Last accessed 2011–10–07.

Loechner, J. 2009b. Kids Motivated by TV and Print to Visit Web. *MediaPost*, January 1. Retrieved from http://www.mediapost.com/publications/?fa=Articles.showArticle&art_aid=97459. Last accessed 2011–10–07.

Madianou, M. 2005. *Mediating the Nation: News, Audiences and the Politics of Identity*. London: UCL Press.

Maggio, F. 2008. The Forest from the TVs. *MediaPost's TVBoard*, December 12. Retrieved from http://www.mediapost.com/publications/index.cfm?fa=Articles.showArticle&art_aid=96609. Last accessed 2011–10–07.

Majority of Americans Believe Television Programming is Getting Worse, Yet They're Watching More Than Ever. 2007. *AOL Television and Associated Press*, September 13. Retrieved from http://ir.aol.com/phoenix.zhtml?c=147895&p=irol-newsArticle&ID=1353995&highlight=. Last accessed 2011–10–07.

Maxwell, R. 1996a. Ethics and Identity in Global Market Research. *Cultural Studies*, 10(2), 218–236.

Maxwell, R. 1996b. Out of Kindness and into Difference: The Values of Global Market Research. *Media, Culture & Society* 18(1), 105–126.

Mermigas, D. 2008. Hulu CEO: More Global Moves Planned for '09. *Media PostNews Online Media Daily*, December 29. Retrieved from http://www.mediapost.com/publications/index.cfm?fa=Articles.showArticle&art_aid=97283. Last accessed 2011–10–07.

Microsoft. 2009. *Europe Logs On: European Internet Trends of Today and Tomorrow*. Retrieved from http://www.scribd.com/doc/14065700/Europe-Logs-On-Microsoft-Study. Last accessed 2011–10–07.

Miller, T. 2009. Cybertarians of the World Unite: You Have Nothing to Lose But Your Tubes! In *The YouTube Reader*, edited by P. Snickars and P. Vondereau. Stockholm: National Library of Sweden, 424–440.

Miller, T. and L. Kim. 2008. Overview: It Isn't TV, It's the 'Real King of the Ring'. *The Essential HBO Reader*, edited by G. R. Edgerton and J. P. Jones. Lexington: University of Kentucky Press, 217–236.

Missika, J.-L. 2006. *La fin de la télévision*. Paris: Seuil.

Morris, J. 2007. Drinking to the Nation: Russian Television Advertising and Cultural Differentiation. *Europe-Asia Studies* 59(8), 1387–1403.

Moses, A. 2009. TV or not TV? *Sydney Morning Herald*, January 9. Retrieved from http://www.smh.com.au/news/technology/biztech/tv-or-not-tv/2009/01/09/1231004268540.html. Last accessed 2011–10–07.

Motion Picture Association of America. 2007. *U.S. Entertainment Industry: 2006 Market Statistics*. Retrieved from <http://www.mpaa.org/resources/7d8f86ee-c14b-42ed-86b1-e478a5824da3.pdf>http://www.mpaa.org/resources/7d8f86ee-c14b-42ed-86b1-e478a5824da3.pdf. Last accessed 2012–02–28.

National Association of Hispanic Journalists. 2006. *Network Brownout Report: The Portrayal of Latinos and Latino Issues on Network Television News, 2005*. Retrieved from http://www.nahj.org/2006/10/network-brownout-reports. Last accessed 2011–10–07.

Neff, J. 2009. Guess Which Medium Is as Effective as Ever: TV. *AdvertisingAge*, February 23. Retrieved from http://adage.com/article/mediaworks/guess-medium-effective-tv/134790. Last accessed 2011–10–07.

Newcomb, H. 2005. Studying Television: Same Questions, Different Contexts. *Cinema Journal*, 45(1), 107–111.

Nielsen Media Research Reports Television's Popularity is Still Growing. 2006. Nielsen Media Research, September 21. Retrieved from http://www.thinktv.

com.au/media/Articles/Nielsen_Media_Reports_TV's_Popularity_Is_Still_Growing.pdf . Last accessed 2011-02-28.

Nordyke, K. 2008. USA Returns as King of Cable. *Hollywood Reporter*, December 24. Retrieved from http://www.hollywoodreporter.com/news/usa-returns-as-king-cable-124853. Last accessed 2011–10–07.

Openwave. 2009. *Privacy Primer: Protecting the Consumer in an Era of Behavioral Marketing*. Redwood City: Openwave.

Pew Research Center. 2005. *Trends 2005*. Retrieved from http://pewresearch.org/pubs/206/trends-2005. Last accessed 2011–10–07.

Pew Research Center for the People & the Press. 2008. *Internet Overtakes Newspapers as News Source*. Retrieved from http://people-press.org/report/479/internet-overtakes-newspapers-as-news-source. Last accessed 2011–10–07.

Postrel, V. 1999. The Pleasures of Persuasion. *Wall Street Journal*, August 2, A18.

Powdermaker, H. 1950. *Hollywood: The Dream Factory: An Anthropologist Looks at the Movie-Makers*. Boston: Little, Brown and Company.

Rash, J. 2009. Oscar Night Ratings Improved, but Far from Golden. *AdvertisingAge*, February 23. Retrieved from http://adage.com/article/mediaworks/oscar-night-ratings-improved-golden/134822. Last accessed 2011–10–07.

Ravel, J. 1999. *The Contested Parterre: Public Theater and French Political Culture 1680–1791*. Ithaca, NY: Cornell University Press.

The Revolution That Wasn't. 2009. *Economist*, April 25, 68, 70.Reynolds, M. 2008, December 24. Nielsen Log: Football, Holiday Fare Rule. *Multichannel News*. Retrieved from http://www.multichannel.com/article/161295-Nielsen_Log_Football_Holiday_Fare_Rule.php. Last accessed 2011–10–07.

Reynolds, M. 2009. CNN, MTV Among Top 10 Most Social Brands: Vitrue Study. *Multichannel News*, February 11. Retrieved from http://www.multichannel.com/article/173998-CNN_MTV_Rank_Among_Top_10_Most_Social_Brands_Vitrue_Study.php?rssid=20059&utm_source=feedburner&utm_medium=feed&utm_campaign=Feed%3A+MultichannelBreakingNews+%28Multichannel+News+-+Breaking+News%29. Last accessed 2011–10–07.

Richardson, D. and M.C. Figueroa. 2005. *Basic Cable Television Industry Research and Corporate Profiles*. Industrial & Labor Relations, Cornell University for the Writers Guild of America, East.

Rincón & Associates. 2004. *Latino Television Study*. National Latino Media Coalition. Retrieved from http://www.rinconassoc.com/nbc4nielsen.pdf. Last accessed 2011–10–07.

Rizkallah, E. G. and N. Y. Razzouk. 2006. TV Viewing Motivations of Arab American Households in the US: An Empirical Perspective. *International Business & Economics Research Journal*, 5(1), 65–74.

Rose, M. M. 2001. Television Industry Profile. *Business.Com*. Last accessed 2002–03–12.

Rosenstiel, T. 2011. Five Myths About the Future of Journalism. *Washington Post*, April 7. Retrieved from http://www.washingtonpost.com/opinions/five-myths-about-the-future-of-journalism/2011/04/05/AF5UxiuC_story.html. Last accessed 2011–10–07.

Sacerdoti, T. 2011. Will the Real Hulu Please Stand Up? *Media Post's Online Video Insider*, March 11. Retrieved from http://www.mediapost.com/publications/?fa=Articles.showArticle&art_aid=146578. Last accessed 2011–10–07.

Shields, M. 2009. Nielsen: Mobile Video Usage Small, but Growing. *AdWeek.com*, January 8. Retrieved from http://www.adweek.com/search/apachesolr_search/Mobile%20Video%20Usage%20Small%2C%20but%20Growing. Last accessed 2011–10–07.

Sky+HD Reveals High Definition Programming Can Help Depression. 2009. *Television PR News*, February 11. Retrieved from http://www.televisionprnews.

com/ae_television_articles/2009/02/skyhd-reveals-high-definition-programming-can-help-depression-86918.htm. Last accessed 2011–10–07.

Smith, N. (n.d.). Know Your Audience. *MarketitOnline*. Retrieved from http://www.marketitonline.co.nz/articles/article/spike-tv-know-your-audience-2825.html. Last accessed 2011–10–07.

Smythe, D. 2004. The Consumer's Stake in Radio and Television. In *Mass Communication and American Social Thought: Key Texts, 1919–1968*, edited by J. D. Peters and P. Simonson. Lanham: Rowman & Littlefield, 318–328.

Spangler, T. 2009. ESPN, Discovery Are Top 'Must-Keep' Cable Nets: Survey. *Multichannel News*, February 19. Retrieved from http://www.multichannel.com/article/174503-ESPN_Discovery_Are_Top_Must_Keep_Cable_Nets_Survey.php. Last accessed 2011–10–07.

Standage, T. 2006. Your Television is Ringing. *Economist*, October 12. Retrieved from http://www.economist.com/node/7995312?story_id=7995312. Last accessed 2011–10–07.

Stiernstedt, F. 2008. Maximizing the Power of Entertainment: The Audience in Contemporary Radio. *The Radio Journal: International Studies in Broadcast & Audio Media*, 6(2–3), 113–127.

Taylor, L. 2006. Culture's Revenge: Laurie Taylor Interviews Stuart Hall. *New Humanist*, 121(2). Retrieved from http://newhumanist.org.uk/960/cultures-revenge-laurie-taylor-interviews-stuart-hall. Last accessed 2011–10–07.

Thomasch, P. 2009. As Advertisers Scrap Deals, TV Faces Cold Spring. *Reuters*, February 20. Retrieved from http://www.reuters.com/article/2009/02/20/us-advertising-idUSTRE51J5PW20090220. Last accessed 2011–10–07.

Turow, J. 2005. Audience Construction and Culture Production: Marketing Surveillance in the Digital Age. *Annals of the American Academy of Political and Social Science*, 597, 103–121.

Umstead, R. T. 2010. Turner: Viewers Watching More TV in 2010. *Multichannel News*, July 12. Retrieved from http://www.multichannel.com/article/454664-Turner_Viewers_Watching_More_TV_In_2010.php. Last accessed 2011–10–07.

Verón, E. 2008. Réquiem para una televisión difunta. *Perfil*, September 28. Retrieved from http://www.gacemail.com.ar/index.php?idnota=10699. Last accessed 2011–10–07.

Watching While You Surf. 2008. *Economist*, June 7, 3–4.

7 Time, Space and Clouds of Information

Data Centre Discourse and the Meaning of Durability

Peter Jakobsson and Fredrik Stiernstedt

This chapter is about data centres: large, dedicated buildings in which interconnected servers are used to store and process digital information on an industrial scale. This information is collected and utilised for commercial or administrative purposes by governments, organisations and companies like Google, Facebook and Microsoft. Arguably, it is the ability (or lack thereof) to collect, store and process information that determines which companies and organisations will dominate the current information economy, as well as the digital media culture of the future. The technologies and business models associated with data centres are commonly referred to as 'cloud computing'—sometimes even hailed as a new computing paradigm (Armbrust 2010)—which has the practical consequence that increasingly more information, as well as the means to process that information, becomes centralised resources in the hands of a few, large actors (Andrejevic 2009).

Data centres are also inscribed in a number of symbolic geographies and staged to perform and reflect informational ideologies and imaginaries. The purpose of this chapter is to analyse the discourses surrounding these buildings—constructed by the data centre companies themselves, as well as governments and municipalities trying to attract data centre investments. What do these discourses tell us about the data centre industry, its business models and the data centre economy's relation to digital culture as a whole?

One central element in the discourse on the so-called *information society* is the celebration of speed and ephemerality: the overcoming of boundaries, the destabilisation of identities and the disappearance of distance (Mosco 2004). Using the example of the data centre industry, however, this chapter tracks an ideological shift within digital culture. The success of cloud computing as a new computing paradigm is dependent on a positive public image of the data centre industry. Since cloud computing implies the centralisation of information—moving information away from end-users towards central warehouses—it is essential that this move is experienced as safe. The success of the computing paradigm is thus dependent on the construction of a discourse in which information is associated not only with speed and ephemerality but also with stability and durability. In the chapter, this discourse is identified and analysed with the help of empirical material collected from government authorities, telecom entrepreneurs and data centre companies, as well as through an analysis of the architecture of the data centres themselves.

Discourse is defined here as a system that structures the way we perceive reality and hence as both an instrument and an effect of power (cf. Foucault 1978/1990). From such a perspective the heterogeneous and multifaceted set of stories about data centres, as well as the symbolic qualities of the data centres themselves (e.g., architecture), take place *within* a certain discourse, while at the same time they are also producing this discourse. Analysing these discourses is thus a way to understand the organisation of power in informational economy and the role played by notions of the durable within this regime. Along these lines, the stories surrounding data centres are interpreted as utterances within an ideological discourse of informational culture that serves political and economic purposes. Data centre discourse is ideological to the extent that it presents the world in a way that naturalises particular (economical) interests, at once promoting and legitimating them (cf. Eagleton 2007). The ideological purpose of the analysed discourse is the main object of analysis in the chapter, but in its final section we will also widen the discussion beyond the strict economic motives of discourses of durability and explore the importance of the dialectics of the ephemeral and durable for understanding contemporary digital culture as a whole.

As an industry, data centres have a long history, but during recent years large companies from the US have increasingly looked abroad for data centre locations (Jaeger et al. 2009). As Internet-based information services become more global, it is advantageous for companies to also develop an infrastructure on a global scale. In this process, for reasons that will be developed subsequently, the Nordic countries (especially Sweden, Finland and Iceland) have come to the fore as suitable locations for this infrastructure, and this is also the region from which we have collected the empirical material for this chapter.

More specifically, we have interviewed representatives from a number of Swedish rural municipalities that are trying to attract large (US) data centre investments.[1] Such a process has been underway for some years, managed by the organisations Invest Sweden and Stockholm Business Alliance. The municipalities are rural, and in many cases the population rates are falling and the unemployment rates increasing. We wanted to get an insight into the process of data centre investment as such, as well as the discourses activated in relation to it: What arguments have been used to attract investments, what hopes and fears are articulated; how is the data centre industry, as well as the activity within the data centres, imagined by the municipalities; and what economic as well as symbolic values are implied in such an investment? The analysis also includes written material from Sweden and the other Nordic countries (PR documents, memos and white papers)—produced by the industry as well as municipalities and government organisations. We have also performed a 'close reading' of the data centre Pionen ('the Peony') in Stockholm, Sweden.

The chapter is structured using three distinct tropes of the 'durable' that have evolved from the empirical material: *geological*, *historical* and *technological time*. In order to stage the data centres as 'durable', 'lasting' and 'stable', they are inscribed in different ways in stories about prehistoric time, for example, the formation of the earth's surface and the organic developments of bedrock. They are also narrativised as part of a 'tradition', inscribed and given a place in history (cf. Ricoeur 1990): political, economic and industrial history, for example. Third, they are placed within *technological time*, whereby technology can be taken to connote the durability of data centres through both technological symbols for control and security, for example, as well as the claims to futurity posed through the technological imaginaries of the data centres themselves. As such, this underlines the fact that the temporal regimes within the discourses of data centres serve the purpose of placing the data centres *in time*, not only to give them a history but also to construct them for the future. The discourses surrounding data centres are intended to convince us that the ephemeral is enduring.

GEOLOGICAL TIME

In a promotional video for the Pionen data centre, located thirty metres under central Stockholm, we learn that it is protected by the Swedish bedrock's unique hardness. This message returns in another short video, uploaded to YouTube by the organisation Invest Sweden, describing Sweden's favorable conditions for data centres:

> Well, I mean the basic foundation for a data center is the stability of the country, right? Stability of the country politically, geographically, climate-wise, there are no natural disasters in Sweden—even the bedrock in Sweden is quite unique in its hardness. (Data Center Pulse 2009b)

This quote captures a peculiar but central theme in data centre discourse: inscribing the question of information storage in a narrative that stretches over enormous periods of time, dating back even to the formation of the earth's surface, and the hard Swedish bedrock. This narrative is replicated in many versions in the material in our study. For example, in another promotional video Dean Nelson, Senior Director of Sun Global Lab, concludes a long statement of praise for the physical surroundings of the Nordic countries—the cold climate that makes cooling efficient, the green energy, the plenitude of water—by saying that 'the fact that you have no natural disasters here . . . certainly it makes a lot of sense to place something in this location' (Stockholm Business Region 2009).

Furthermore, such a story is reproduced by representatives of municipalities seeking to attract international data centre investments. One of our interviewees explains that he has never experienced a more thorough and demanding evaluation process than those instigated by these international companies. The data the companies in question have requested concern 'climate conditions, how the winds blow, and the risk of earthquakes', as reported by one of the informants. Another representative of one of the municipalities talks about the following issues:

> Earthquakes, volcanic eruptions, floods. For these companies this is of course an international question: 'Where in the world are we to place our facilities'? 'Where should we locate these nodes, the production centrals'? And then questions of, for example, movements in the earth's crust become important . . . In Europe there are different conditions in different parts. Sweden and the Nordic countries are extremely unique in this respect.

As our respondents rightly point out, Sweden and the Nordic countries have unusually stable geological conditions, both due to the composition of the bedrock and because the Scandinavian Peninsula is located far from the edges of the Eurasian tectonic plate. But this should not only be understood as a matter of fact, since as we will see, this specific geography also has to be represented in some form and placed within a coherent narrative.

The demands raised by the Internet companies in question push the cities and their representatives to promote their specific locations as secure and to stage them as durable geographies, a process in which 'one learns a lot about things one doesn't normally think about', according to one of our interviewees. In the communication between the Internet companies and their potential partners, new dimensions of the geographies are highlighted and put together in order to accomplish such an image of stability. The process is described by the informants as one of hard labour, in which 'really thorough Excel sheets' with 'hundreds of variables' have to be answered.

> I would say that I personally have put 200 hours into this. And I have a staff member who has done at least as much. And then there are about 15–20 people involved in this, so I would say that all in all the work time for filling in such forms is close to 1,000 hours. And then I'm only thinking of one single case.

Stability on a geological timescale is thus not only a 'natural' quality of the Swedish landscape, but also largely something that is staged and constructed in the application process.

The architect of the Pionen data centre also highlights how important stability is in the design and presentation of the data centre's infrastructure. In the design of Pionen he wanted to create a contrast between the server rooms—where the human is 'a stranger', in the way the walls are laid bare, exposing the 'rockiness' in the walls—and the rooms that are furnished and adapted to a human scale, according to the architecht, Albert France-Lanord. The servers and the information stored in them are thus inscribed in a geological timescale and in a spatial narrative of stability. In contrast to this, the work spaces emphasise familiar temporal routines and organic materials that make the stable and unchangeable environment of the server halls stand out even more. In this way, information is staged as overcoming the shorter human time scale and the inevitable entropy of death.

HISTORICAL TIME

In a report by data centre consultant John Rath on principles for data centre site selection, one of the central dimensions is said to be 'political stability and security, family life, community life, job security, political freedom and gender equality' (Rath, 2010). A report from *Global Facilities Development* distinguishes between what they call natural and unnatural disasters, the latter emanating from political and social factors (Global Facilities Development 2010). The importance of social and political stability is also furthered by the interests that seek to attract data centre investments, and is hence recurrent in our material. The organization Invest Sweden holds as one of their 'top ten reasons' for establishing a data centre in Sweden the particular political, social and physical stability of the country (Invest Sweden 2009). In the election campaign of 2010, the Swedish Christian Democratic Party (now in government) furthermore made it one of their areas of concern to create even better political and policy frameworks for data centre establishments, since

> Sweden should have good preconditions to be world leading in attracting new investments in server halls. *Political stability*, cold climate, cheap energy and a highly educated population contribute to making our country especially well suited for growth within this sector. (Kristdemokraterna 2010)

Political stability, in general terms, is of course always an issue when deciding on where to locate big investments, and plays a part in the site selection of any major company. In relation to this specific industry, however, as discussed in the previous section, stability has a very specific meaning. Political stability is thought of not only in strict economic terms, but also as a safeguard against threats to the durability of the data preserved within server farms—as a counter-agent against the inevitable entropy of information. The stories the towns and municipalities use to attract these kinds of investments, as in the preceding section, thus intentionally seek to include and perform such elements. This is the case, for example, in Arboga. The representative of business establishments tells of the strategy adopted by the municipality:

> We like to be understood as a cultural city and a town with a history . . . That is, you could say, the brand of Arboga . . . So when we have meetings or gatherings of various kinds here . . . we always try to design the meeting so that innovation and digital technology are highlighted next to history and culture.

The medieval town of Arboga wants to be perceived as both a symbol of the historic, old and durable—being one of the oldest towns in the country—and simultaneously as a town that is modern and progressive. And in this case these ideals and aspirations may converge, since Arboga's fame comes from the fact that it hosted the first Swedish parliament in 1435.

Similar stories of political and social stability are being constructed in the other Nordic countries. For example, Thordur Hilmarsson, a managing director at the nonprofit Invest in Iceland Agency (IIA), seeks data centre investments on the basis of Iceland's 'political stability and political integrity' (Trujillo 2007). Since Iceland does not have the same geological preconditions as the other Nordic countries, it is natural that they need to emphasise other aspects of their country:

> Iceland has no military and has never actively engaged in war with other nations. The country is a founding member of NATO and has had a defense agreement with the United States since 1951. Corruption is minimal, political risk is very low, and the crime rate in Iceland is one of the lowest in the world. (Invest in Iceland 2010)

The specificities of the social history and tradition, however, can also become problems. As one of our informants states, the companies he has been in contact with 'are not informed about the unique Swedish legislation, about our high degree of transparency' (Helmer Larsson, interview). This shows that there also might be tension or conflict between the sought-after political and social safety, which is produced through longstanding democratic traditions and high levels of transparency, and the other elements needed (i.e., security, secrecy, anonymity) in the production and manifestation of the durable in informational culture.

A clip, produced and uploaded to YouTube by the operators of the Pionen data centre, starts by showing the massive early 20th century, romanticist church in red granite, located at the top of a rock above the data centre.[2] The church bells are ringing and continue to ring as the image of the church fades out and the spectator finds himself inside the underground data centre. What does this alignment between the church and the data centre mean? They are—coincidentally—located at the same place, but it is no coincidence that the promotional film chooses to underline this fact. The establishing shot of the church accomplishes two things. First it juxtaposes, rather than aligns, the two structures—using the church's connotations of tradition and ritual as a contrast to the newness of the data centre. The church—both as an institution and as a building—and the regime of communication are thus used to stand in opposition to the ephemerality of the digital. Yet in order to make this juxtaposition there must also be some kind of affinity or similarity between the two, which is also made clear in the clip when the audio track connects the two buildings through the sound of the ringing church bells. We can understand the affinity created in the clip by following the argument in the introduction to the edited volume *Media Houses* (Ericson and Riegert 2010). In Victor Hugo's *The Hunchback of Notre Dame* it is proclaimed that the printed word and the printing press will vanquish the church's monopoly on knowledge. This will take place since the printed word will extend the lifespan of everything printed. No matter how effective the church is in persisting through time, the printed word can override its importance and influence by scattering knowledge 'to the four winds, and occupy[ing] all points of air and space at once' (Ericson and Riegert 2010: 4). What we see in the clip is how the data centre is inserted into a narrative in which a new and more effective system of communication is taking over the position of an outdated system.

Part of the historical narratives are also the concrete *industrial* histories in which the data centres inscribe themselves, which are also important resources in producing narratives of durability and stability. Such a history often plays a dominant role in the narratives surrounding the installations. Physically, the data centres often inherit locations previously used for 'heavy' industry, such as aluminum smelters or steel mills, since they consume a great deal of energy. Therefore, the relationship between and discourses surrounding the 'old' and 'new' industrial regimes are actualised in almost every new data centre placement.

> In our region we have had a lot of heavy industry. Mines. Steel industry. Paper mills. Sawmills . . . And we need to renew the structure of our entire trade and industry sector here.

And in the largest Nordic data centre investment, in the Finish town of Hamina, representatives of the town are very clear in how they see the industrial history of the area, perpetuated and continued within the 'new' industry, in this case represented by Google. In the following passage, from their website, they connect not only the industrial but also the media histories of the 20th and 21st centuries, as well as relating one of the portal figures of modernism (in architecture) to the arguable 'post-modernity' of the Silicon Valley company. The following quote connects all these in a story about the old Hamina paper mill.

> The symbolism is striking: Where once paper machines produced colossal rolls of newsprint, ready to carry printed information to people's homes, another kind of machine will soon connect some of the same people with electronic information in the form of internet pages like the one you're reading now.
>
> That's not all: In this old port town (the name Hamina is derived from "hamn", the Swedish word for "harbour"), where goods used to arrive and leave via the docks, a waterfront building will now receive and redistribute information.
>
> And to top it all off, two parts of the Summa mill were designed by world-famous Finnish architect Alvar Aalto. (Marten 2009)

TECHNOLOGICAL TIME

Geological and historical time stretch over long time periods, spanning many generations. Technological time, however, as we define it in this chapter, is shorter. Although different industries and sectors operate within different time scales—the time before obsolescence for a generation of airliners is longer than for video game consoles—when it comes to the technologies we are dealing with in this chapter, the time span is much shorter than the geological or historical time span. In this section of the chapter, we look at the Pionen data centre located in Stockholm at the top of 'the White Mountain'. This is a direct translation of the Swedish name (*Vita Bergen*), but it is also the name of a science-fiction novel from 1967, written by John Christopher. And as we will see, science fiction also plays a major part in the staging of this location.

A commercial brochure for Pionen asks, 'How many megatons can your server take?' (Bahnhof n.d.). Pionen was built during the Cold War and is constructed to withstand nuclear warfare, once acting as an emergency central in the case of war. The director of the facility says in a promotional video clip that they are aiming for 'ultra security' and that this is guaranteed by the facility's military past (Bahnhof 2009b).

> As a matter of fact the mountain is self-sufficient. If Stockholm were hit by natural disaster or terrorist attack that affected the power supply or data and telecommunications, Pionen could function as a national communication node for weeks (Bahnhof n.d.)

In a Japanese TV show about Pionen we are also shown that the doors leading into the facility, located thirty metres underground, are seventy centimetres thick and made of solid steel (Bahnhof 2009a). According to the architect Albert France-Lanord, they were actually made this thick not for functionality but because of the symbolic value of very thick steel.

The site features many such references to control rooms, bunkers and hideouts of super villains and power-crazed scientists from the world of (Cold War) science fiction. Pionen's architect says he took a great deal of inspiration from the set designs of Ken Adams—who designed the sets of *Dr. Strangelove* and the James Bond films *Dr. No* and *Goldfinger* (Saieh 2008). Pionen's director also associates the layout with superhero narratives, saying he first wanted to build an elevator with which you could access the data centre via the top of the mountain—and claims he got the inspiration to do so from the 'Batcave', the home of action superhero Batman (Data Center Pulse 2009a). In the Batman universe, the cave functions as a sanctuary: It is a place where Batman can safeguard himself from both his private life and the super villains hunting him in his role as the Caped Crusader. In this way, Pionen is taking part in discourses and narratives about security and control that are key in many discussions about data centres.

But control and security are also accompanied by discourses that in a more general sense have to do with time, history and futurity. The material imaginaries of the data centre are not inscribed in a straightforwardly modern or modernist project, understood as a 'surge for the new'. Rather, the facilities are staged as a dialogue between the 'old' and 'new' through their references to sci-fi nostalgia and the conspicuous architecture of the Cold War, coupled with the most advanced computer technology and contemporary symbols of

Figure 7.1 Pionen data centre. Photo by Åke E:son Lindman. Reproduced with permission.

environmental sustainability, showing off the data centre as a 'green building'. This retro-futuristic décor arguably stages the data centre as a place that will remain: a building that is already set for the future.

In using 'mediatized memories of the cinescapes of science fiction' (Lagerkvist 2010: 226), however, this staging is somewhat contradictory. On the one hand, the data centre is made to resemble the hideout of a super villain or mad-genius scientist of Cold War science fiction movies. On the other hand, the data centre plays with symbolic material from the superhero discourse, for example, in the explicit references to the Batcave, mentioned already. The architecture hence alludes to both the contenders and defenders of the existing order. The way these mediatised memories blend in with the architecture and interior design of Pionen also tells of their shared origins, and the fact that they are the two ways in which what Paul N. Edwards (1996) has called the 'closed world' of Cold War culture was symbolically acted out in popular culture. The computer centre within the bunker-cave in Stockholm is in a very concrete way a 'closed world': a sealed off, claustrophobic space. As such, the mediatised memories staged within the data centre can be perceived as a simulacrum that is intended to symbolically reassert the (however illusionary) stability and stasis of the recent—yet strangely foreign—Cold War past. At the same time, such mediatised, science fictional architecture can also play the part of invoking what Amanda Lagerkvist has called a 'retromodernity', using 'scenery' that has long been regarded as images of the future. As such, Pionen's architecture creates a certain retroactive sense of futurity through reigniting the thrust of a past understanding of the future (e.g., what the year 2000 looked like in the 1960s).

The intention of this discourse of futurity in the architecture is arguably to stage Pionen in a narrative of the durable—built for atomic warfare, constructed for eternity. Even the name—Pionen or the Peony—gives a hint of such a pretension, as the flower is well known for its 'conservatism': Once planted, the peony likes to be left alone and punishes those who try to move it by not flowering again for several years. A final indication that this is also the intention of the operators themselves is provided by the architect, who quotes the science fiction film *Silent Running* as *the* central source of inspiration for the design of the data centre (Saieh 2008). *Silent Running* is a cult movie from the 1970s about a fleet of space ships carrying precious cargo consisting of the last remaining specimens of Planet Earth's plants and wildlife—traveling through space like Noah's Ark or, if you will, an archive over the earth that once was; but in any case as a time capsule, encapsulating, preserving and carrying its cargo forward into the future.

In one way, the data centre as a spaceship/ark hence becomes a utopian counterpart to the ultimately tragic narrative of history and modernity that Walter Benjamin provides in his famous ninth thesis on history. The *Angeles Novus*, the angel of history, is carried by the storm from paradise, unable to halt his movement or 'awaken the dead and make whole what has been smashed' in the course of progress and modernity (Benjamin 1968:257). What is expressed through the materiality and symbolic dimensions of the Pionen data centre is a similar, yet different, philosophy: The data centre as spaceship is carried away by the storm of progress—flying through space towards 'the future'—taking part in the development towards modernisation and informationalisation. But at the same time it is an ark that gathers, collects, stores, preserves and encapsulates the 'debris' created by that same storm. It is—perhaps—a vain attempt to simultaneously move through time and bring along the remains of this movement: the infinite 'cloud' of bills, e-mails, and PMs; weather and high-school football reports; bank statements; photos, videos, television programs; book manuscripts and military secrets. It is hence not as much an archive as a collection; it is not a systematised and systematising attempt to create history, but rather a simultaneously more haphazard and grand project—to make the ephemeral endure.

THE DIALECTICS OF DURABILITY AND EPHEMERALITY

The chapter has explored what appears as the binary opposition between the ephemeral and the durable informational culture. We have identified three different but related narratives that operate on three time scales: geological, historical and technological time. But what is the purpose of these narratives? We have already pointed to needs of the data industries to sell security and stability, in order to attract more of our data into their data centres and server farms. In this they are also collaborating with the countries and municipalities that seek investment from these global companies. As a concluding discussion, however, we would like to go beyond these strict economic motives and discuss a more deeply rooted ambivalence haunting digital culture, related to the dichotomy between ephemerality and durability.

The opposition between ephemerality and durability is present in narratives about the information society, but also in narratives about modernity and post-modernity. As captured by Karl Marx in the famous phrase 'all that is solid melts into air', modernity has been understood from the outset as a movement that uproots tradition and compulsively produces the new (Berman 1983). For contemporary theorists such as Frederic Jameson (1991), this impulse has been radicalised to the point at which *time* itself is annihilated and history eradicated—leaving *space* as the relevant category with which to think about culture and society. Doreen Massey has furthered this argument, but has also pointed to its internal contradictions. Digital and electronic communications are instantaneous, and consequently caught up within a belief of the eternal now, but, on the other hand, they also bridge distances in a way that could make us think that it is rather space—or both time and space—that has lost its meaning (Massey 2005). The established narrative on the information society is an extrapolation from its most fundamental cultural form—digital information—to society as such. The supposed ephemerality of information is claimed to undermine the fabric of social organisation. To conclude, we would like to add a few more notes in regard to this, going beyond the strict ideological needs of the information industries and touching upon questions of everyday experience, cultural continuity and social structures.

First of all, what do the analysed narratives tell us about the meaning of durability from the perspective of personal identity and everyday experience? Manuel Castells (1997), among others (e.g., Huyssen 1995; Bauman 2001), has commented on the relationship between the speed with which information travels around the globe and an increased need to find some kind of anchoring in the local, or in tradition and history. Some of the narratives in this chapter connect to such a relationship between macro developments and personal experiences of rootlessness. If we formerly relied on diaries, photo albums, books and various building sites and architectural landmarks to find such an existential anchoring, we can speculate that there is a similar need to know that our digital lives are equally durable and stable. In this, the material information infrastructure plays some role. There is a certain experience of psychological relief in witnessing the steel doors and underground bunkers that are used to store some of our personal information.

Another interesting aspect of the material in our study is that the same anxiety that is experienced on a personal or communal level is also present within the world of business and finance. One example was how the search company Google, whose search engine indexes the ever-expanding information on the Internet, is also involved in projects attempting to preserve things of more lasting cultural value, such as the works of William Shakespeare. From this perspective we might understand the data centre as trying to occupy the same place in our culture as the library has traditionally done: 'an institution that has enabled communication between people over time and space, and has been an indispensable tool for man's development as a social and cultural being' (Bolin 2009: 13). The emphasis on durability in the construction of data centres is thus a part of a broader wish for cultural continuity, and to avoid the fate of the most famous library of all times—the library of Alexandria, which according to legend was burned to the ground in 48 BC.

A third dimension of the discourse of durability in informational culture serves as a counter-story to the increasing demands and challenges of globalisation. Forces of mobility and flexibility—and the increasing movements and destabilisations of a postindustrial and neoliberal economic regime—create not only (and not even mainly) economic growth but also uneven development and economic insecurity. Inscribing the 'new economy' within the familiar story of the industrial society is hence an ideological move that serves the purpose of preventing unwanted social distress. The image created by the discourse of durability is that the data centres have the potential to become 'a new basic industry' or 'the factories of tomorrow', as one of our informants said. It seems to be beside the point that these investments in reality do not deliver any significant number of new jobs or that the economic surplus will move abroad. The 'brand value' of these facilities, the fact that they represent a durability that will carry them forward into the future, and their image of newness and futurity seem to be enough.

As a final conclusion we thus wish to again stress the dialectical movement between the ephemeral and the durable that is visible in the data centres as symbolic artefacts and in the discourses that surround them. Our argument is that the need for narratives of durability, history and tradition are formed in relation to those on the ephemerality of information and its spatial reach. What we have presented is not an alternative story to the more familiar one of ephemerality, but rather something that is intimately intertwined with it. The data centre is a potent resource in providing the necessary counterpoint for such a story, but as we have seen it is a counter-point that needs to be crafted very carefully. The care with which this crafting is undertaken points to both the perceived ephemeral nature of information and the cultural and economic need to provide it with durability.

NOTES

1. The interviewed were the Heads of Establishment in Östersund (Magnus Lindgren), Luleå and Piteå (Matz Engman), Västerås (Helmer Larsson), Ludvika (Lars-Åke Josefsson) and Arboga (Göran Dahlén), as well as the Heads of ICT Investment, Stockholm Business Alliance (Torbjörn Bengtsson). We have also interviewed the architect of the centre, Albert France-Lanord. All interviews were made in October 2010.
2. http://www.youtube.com/watch?v=wn8pz1HLYp8. Last accessed 2010–10–31.

REFERENCES

Andrejevic, M. 2009. Cloud Control. Paper presented to the Popular Communication Division of International Communication Association conference, Chicago, May. Armbrust, M., A. Fox, R. Griffith, A. Joseph, R. Katz, A. Kowinski, G. Lee, D. Patterson, A. Rabkin, I. Stoica, and M. Zaharis. 2010. A View of Cloud Computing. *Communications of the ACM*, 53(4), 50–58.

Bahnhof. n.d. *Bahnhof Pionen*. Retrieved from http://www.bahnhof.se/data/filestorage/prodblad/pionen_final_neu.pdf. Last accessed 2010–12–16.

Bahnhof. 2009a. *Bahnhof's Computer Center in Japanese TV*. Retrieved from http://www.youtube.com/watch?v=VGI-hJRS9Ec. Last accessed 2010–12–16.

Bahnhof. 2009b. *Pionen—Bahnhof's Computer Facility Movie1*. Retrieved from http://www.youtube.com/watch?v=wn8pz1HLYp8. Last accessed 2010–12–16.

Bauman, Z. 2001. *Community: Seeking Safety in an Insecure World*. Cambridge: Polity.

Benjamin, W. 1968. *Illuminations*. New York : Schocken Books.

Berman, M. 1983. *All that is Solid Melts into Air: The Experience of Modernity*. London: Verso.

Bolin, G. 2009. Kulturinstitution, medium, arkiv: Bibliotekets symboliska betydelse i kultur och samhälle. In *En bok om böcker och bibliotek*, edited by E. Jansson. Huddinge: Södertörn University, 13–20.

Castells, M. 1997. *The Information Age: Economy, Society, and Culture.* Vol. 2: *The Power of Identity.* Malden, MA: Blackwell.

Data Center Pulse. 2009a. *Episode 11—White Mountain, Fit For A James Bond Villain!* Retrieved from http://www.youtube.com/watch?v=qwlATf9xse4. Last accessed 2010–12–16.

Eagleton, T. 2007. *Ideology. An Introduction.* London: Verso.

Edwards, P.N. 1996. *The Closed World. Computers and the Politics of Discourse in Cold War America.* Cambridge, MA: MIT Press.

Ericson, S., K. Riegert and P. Åker. 2010. Introduction. In *Media Houses: Architecture, Media and the Production of Centrality,* edited by S. Ericson and K. Riegert. New York: Peter Lang, 1–18.

Foucault, M. 1978/1990. *The History of Sexuality.* Vol 1: *The Will to Knowledge.* Harmondsworth: Penguin.

Global Facilities Development. 2010. *Unnatural Disasters.* Retrieved from http://datacenterforlease.net/unnaturaldisaster.html. Last accessed 2010–12–16.

Google. 2008. Bringing *History Online, One Newspaper at a Time.* Retrieved from http://googleblog.blogspot.com/2008/09/bringing-history-online-one-newspaper.html. Last accessed 2010–12–19.

Google. 2010. *About Google Books.* Retrieved from http://books.google.com/googlebooks/thoughts.html. Last accessed 2010–10–30.

Huyssen, A. 1995. *Twilight Memories: Marking Time in a Culture of Amnesia.* New York: Routledge.

Invest in Iceland. 2010. *Iceland: The Coolest Location for Data Centers.* Retrieved from http://www.invest.is/investment-opportunities/data-centers-in-iceland/. Last accessed 2010–12–16.

Invest Sweden. 2009. *Why Sweden?* Retrieved from http://www.investsweden.se/North-America/Industries/ICT/Data-centers/Why-Sweden/. Last accessed 2010–12–16.

Jaeger, P. T., J. Lin, J. L. Grimes and S. N. Simmons. 2009. Where is the Cloud? Geography, Economics, Environment, and Jurisdiction in Cloud Computing. *First Monday,* 14(5).

Jameson, F. 1991. *Postmodernism, or, the Cultural Logic of Late Capitalism.* London: Verso.

Kristdemokraterna. 2010. *Kommunikationer och IT.* Retrieved from http://www.kristdemokraterna.se/VarPolitik/Politikomraden/KommunikationerOchIT. Last accessed 2010–12–16.

Lagerkvist, A. 2010. The Future is Here. Media, Memory and Futurity in Shanghai. *Space and Culture,* 13(3), 220–238.

Marten, P. 2009. Google Finds Itself in Finland. *This Is Finland.* Retrieved from http://finland.fi/public/default.aspx?contentid=163306&nodeid=41805&culture=en-us. Last accessed 2010–12–16.

Massey, D. 2005. *For Space.* London: Sage.

Mosco, V. 2004. *The Digital Sublime. Myth, Power and Cyberspace.* Cambridge, MA: MIT Press.

Rath, J. 2010. Data Center Site Selection. *Datacenterlinks.com.* Retrieved from http://rath-family.com/rc/DC_Site_Selection.pdf. Last accessed 2011–09–27.

Ricoeur, P. 1990. *Time and Narrative.* Vol. 1. Chicago: The University of Chicago Press.

Saieh, N. 2008. Pionen—White Mountain/Albert France-Lanord Architects. *Archdaily.com.* Retrieved from http://www.archdaily.com/9257/pionen-%E2%80%93-white-mountain-albert-france-lanord-architects/. Last accessed 2010–10–31.

Stockholm Business Region. 2009. *Stockholm as a Data Center Location.* Retrieved from http://www.youtube.com/watch?v=eQbA9eYaZKU. Last accessed 2010–12–16.

Trujillo, A. 2007. Data Center Site Selection Goes International. *SearchDataCenter.com*. Retrieved from http://searchdatacenter.techtarget.com/news/1284739/Data-center-site-selection-goes-international. Last accessed 2011–09–27.

8 Search Engines in Practice

Structure and Culture in Technical Development

Elizabeth Van Couvering

This chapter uses the case of the Internet search engine to investigate the connection between local structures of technological development and larger industrial and cultural systems in which that development is typically situated. It examines the accounts given by search engine engineers of how they produce search engine results and compares their accounts to the historical development of the industry, using insights derived from structuration theory (Giddens 1984) and from subsequent critiques of the theory.

Search engines, such as Google, Yahoo! and Microsoft's Bing, are an essential element of the Internet. The vast majority of web users use them; they have consistently ranked as the most highly visited sites in every country where statistics are collected (although beginning to see competition on this point from social networking services such as Facebook); and they are multi-billion dollar companies which, in turn, funnel traffic and therefore sales to many other businesses large and small across the World Wide Web.

This chapter investigates how search engine bias comes to be embedded into both local practices and larger systems. *Search engine bias*, as I use the term here, consists both of systematic bias towards classes of websites and specific exclusion or censorship of particular websites. For many years, scholars have been reporting that search engines give a particular slant to the web (e.g., Introna and Nissenbaum 2000). The results retrieved for a given query represent only a portion of the potentially searchable web and over-represent popular sites, commercial sites and American sites. (Bergman 2001; Kleinberg 1998; Kleinberg and Lawrence 2001; Lawrence and Giles 1999; Vaughan 2004; Vaughan and Thelwall 2004). The bias of search has been further examined by, among others, Hargittai (2000), Moshowitz and Kawaguchi (2002, 2005), Goldman (2006) and Diaz (2008). Nor is it simply the companies which produce search engines which are implicated in the final shape of the results: Machill et al. (2003) and Rogers (2000) also highlight the important of other actors, including search engine marketers, in creating search engine results as a commercial space. Nevertheless, the search engine production companies and the technology they produce must be the major focus of any investigation of search engine bias.

Because of the central position of search engines in the Internet and the increasing reliance by both public and private organisation on the delivery of essential services online, the understanding of the mechanisms of search engine bias is of general interest. However, studies which focus on technical development often fail to address wider issues of context in which the development takes place. This chapter explicitly connects the industrial structure and history of the Internet search engine to the practice of its producers. The empirical evidence for the present study has been previously set out (Van Couvering 2007, 2008a, 2008b) and rests on in-depth interviews with senior search engine engineers carried out in 2003–2004 and on a historical analysis of primary and secondary source material.

The chapter begins with a discussion of search engine practice as viewed through the lens of structuration theory; it then moves on to link that practice to wider systems and structures in society using cultural critiques of structuration theory as a guide; this system then is related to ongoing historical developments within economics and technology which highlight the relation of rhetorical practice to the development of economic resources. The final section considers how the culturally constructed practice of search relates to public concerns about search engine quality that remain outside the cultural schemas used within the industry.

THE PRACTICE OF SEARCH

It is a premise of the social study of technology that the various technologies we see around us are things that must be explained with reference to social processes and social actors. This method of investigating, which focuses on the historical, the contingent and the social, challenges an earlier paradigm that treated technology as a 'black box', an object following a trajectory that was best analysed by engineering rather than social science methodologies.

In 1992, Orlikowski set out a 'model for analyzing the nature and role of technology in organizations' (Orlikowski 1992: 398) which she called the 'structurational model of technology', drawing on structuration theory (Giddens 1984). She later expanded the model (Orlikowski 2000) to develop a more sensitive idea of user interactions, which she calls the 'enactment' model or the technology-in-practice model. In these articles, she reframes the role of technology 'in terms of a mutual interaction between human agents and technology, and hence as both structural and socially constructed' (Orlikowski 1992: 403). Her second article (Orlikowski 2000) proposes what she calls a 'practice lens' for studying technology. In this, she emphasises the recursive and repetitive nature of technology use as the principle way in which structures become enacted, based on the properties of the technological artefact and the various capabilities, interpretative schemes, and norms of the users and of their situation. This further development of the technological structuration model allows for a better engagement with technologies such as search engines which are difficult to consider stabilised and through which different kinds of structures may be enacted.

Orlikowski's model can be applied to the analysis of search engines as they are practiced by search engine engineers, as shown in Figure 8.1. At the very bottom of the figure is the box containing a description of the ongoing, situated use of technology in the search engine production context. This describes the everyday, routine actions of those engineers involved in search production. Interviews with search engineers suggested a daily routine of relevance judgements, weighting, tuning, developing the index and combating spammers, as well as adapting search to new formats such as video in order to improve the twin measures of relevance and customer satisfaction (Van Couvering 2007).

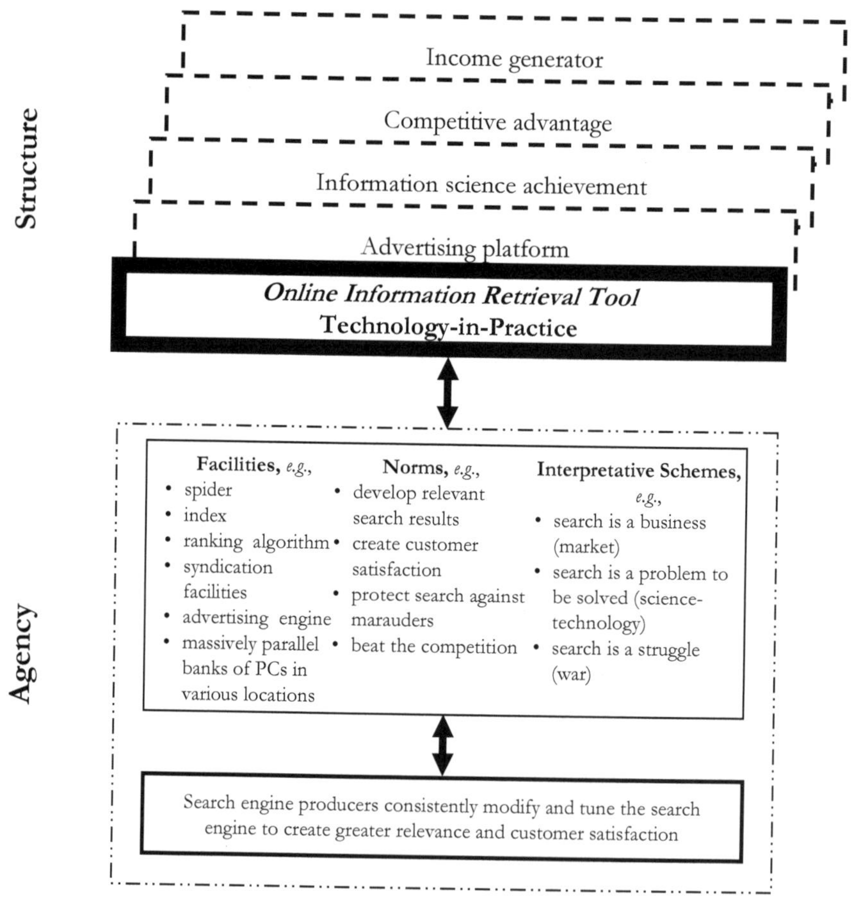

Figure 8.1 Online information retrieval technology-in-practice enacted by search engine engineers. Figure by author.

Moving up the figure, the middle box refers to the modalities of interaction—the *interpretative schemes*, *norms* and *facilities*. Orlikowski has discussed these in terms of assumptions and knowledge about technology, protocols and etiquette related to technology, and hardware and software, respectively. Analysis of interviews conducted earlier (Van Couvering 2007) showed the major schemes as those of technology as a business (termed the *market* schema), technology as a problem to be solved (the *science and technology* schema) and technology as a struggle (the *war* schema). The norms observed are related to these—the creation of relevant searches that boost customer satisfaction and the emphasis on secrecy to protect search engine results and win against the competition. Finally, the facilities which help to bring these norms and interpretative schemes to fruition include the algorithms which run the search as well as syndication and advertising algorithms, and extremely large banks of computers and electrical power situated worldwide.

The top box of Figure 8.1 describes the structures that are instantiated in the routine situated use of search engine technology by search engineers. In this context, a *structure* is a rule-resource set that exists because it is mobilised by agents in their ordinary practice, and it is precisely this kind of structure to which Orlikowki refers by the term "technology-in-practice" (Orlikowski 2000). The situated action which motivates and is motivated by the modalities discussed above, can be understood as enacting the search engine primarily as a tool for online information retrieval and, recursively, it is in pursuit of better online information retrieval (e.g., adding the ability to search for video, or search within local geographic boundaries) that much of the day-to-day work of the search engine engineer is structured. Other technologies-in-practice, however, are also instantiated and a consideration of the evidence suggests that these include, at least, search as an advertising platform, search as an information science achievement, and search as a competitive advantage for the corporation. Finally, the search engine as a technology is an income generator—the source of a search engineer's livelihood.

Having discussed how the technology-in-practice model can be applied to search engine producers, the preceding analysis still presents us with a number of problems. First, in this model the different groups, that is to say, the search engine producers and any potential other groups, for example, search engine marketers,[1] are presented as separate, whereas they would interact along multiple dimensions, sometimes conflicting, sometimes co-operating and sometimes operating relatively independently. Second, although the model captures a process, it gives little hint as to the dynamic elements of the process—in a sense, it implies that the structuration process described is self-sustaining. Third, the relation of the structuration processes captured in the model to wider social systems remains unclear. This relates to the partial use of structuration theory upon which Orlikowski builds her technology-in-practice model, which is, in turn, related to her

the part of the speaker, and that different 'interpretative repertoires' have different functions for speakers and will appear in different contexts. The strategic use of interpretative repertoires fits much more closely with the empirical data in this study than the unified technological frame approach of either Bijker or Orlikowski and Gash.

According to earlier research (Van Couvering 2007), search producers found the market schema is especially productive when it comes to mobilising resources within the pre-existing structure of a business. Nevertheless, producers did not seem to feel that they are experts in the market. They might be engineers, researchers, or designers and they might become managers or even business owners over the years. However, they did not appear to identify themselves as business people and when working on search engines they seemed to feel that their ability to affect major features of search was circumscribed as soon as their companies began to make significant amounts of money from the search. They acted discursively to reclaim their abilities to control search by claiming expertise on an objective measure of search quality, which they linked to customer satisfaction and therefore revenue.

Thus, although search producers, like all other cultural actors, have a multiplicity of transposable schemas at their disposal (examples might include art, or public service, or family, or many others), the implication is that these schemas are not a simple matter of choice—they have real, material consequences for their users.

This interpretation of structuration theory indicates that technological schemas and associated norms will have an effect how resources are directed within society, helping to maintain old structures or create new ones. In the case of search production, it is clear that a considerable amount of resource is dedicated to maintaining search quality in the form of relevance and of customer satisfaction. And as Sewell maintains, the accumulation of resources in turn *validates* schemas and ensures their continued use. As he puts it, 'Schemas not empowered or regenerated by resources would eventually be abandoned and forgotten, just as resources without cultural schemas to direct their use would eventually dissipate and decay' (Sewell 1992: 13). In this way it seems inescapable that pre-existing systems will profoundly influence which kinds of technological schemas and norms become widespread and powerful.

THE SEARCH SYSTEM

The intertwining of symbolic and material culture through practices, outlined in the cultural approach to technological structuration in the previous section, has implications for understanding just how the everyday decisions and actions of search producers and search optimisers relate to the historical development of search engines. In an earlier study of search history (Van

Couvering 2008b), I examined an industry less than two decades old, which had within ten years of its inception begun to show many features of established media industries including an oligarchic, consolidating structure, based on advertising and controlled from a base in the US. This same industry also contained many new elements such as virtual integration of content producers and the change of the essential commodity sold from audiences to traffic. The analysis conducted previously still gives little insight into how such a system could have been built up so quickly, apart from a sense that it must be related in some way to pre-existing norms, interpretative schemes and facilities, which themselves lead to certain technologies-in-practice (like search-as-information-retrieval-tool and search-as-marketing-platform). In turn these technologies-in-practice are related in some way to the larger system of the search industry, itself a part of global capitalism. My earlier study also showed that actors within search industries used particular technological schemas strategically to mobilise resources. As the previous section discussed, the very existence of resources is also in some sense based upon their construction through practices. This section uses that insight to relate the structural development of search to cultural practices.

The history of search must be considered in the context of the dot-com boom, the explosion of investment into Internet technology which peaked in 2000 and was shortly followed by the dot-com bust, a dramatic fall in the stock market and particularly in the value of Internet stocks. Perez (2002), in her historical analysis of the similar booms and busts that have occurred regularly since the 1770s with the start of the Industrial Revolution in England, discusses the linkage between technological, cultural and financial elements that characterise these economic cycles or long waves.[3] Each cycle, Perez suggests, follows the sequence *technological revolution–financial bubble–collapse–golden age–political unrest* and is driven by developments in technology leading to major increases in productive capacity, and also by associated speculation by the financial markets into firms using those technologies. The *technological revolution* is defined as 'a powerful and highly visible cluster of new and dynamic technologies, products, and industries, capable of bringing about an upheaval in the whole fabric of the economy and of propelling a long-term upsurge of development' (Perez 2002: 8). Importantly, associated with this revolution is an associated change in practice and meaning, which Perez calls the *techno-economic paradigm*, which she defines as 'a best-practice model made up on a set of all-pervasive generic technological and organisational principles . . . When generally adopted, these principles become the common-sense basis for organising any activity and structuring any institution' (Perez 2002: 15). Although Perez does not discuss the idea of practice, this 'common sense' that Perez refers to can be related to the idea of technology-in-practice, if it is interpreted as a new schema of the kind discussed in the previous section; in other words, as a cultural vehicle which enables proponents to marshal resources that have been transformed by the technical revolution.

This new techno-economic paradigm spreads from the initial industry where it first appears throughout the whole economy, and then through society as a whole. For example, Perez describes the 'mass production' paradigm as follows:

> Until the 1980s, the prevalent organization was the one that served as the optimal framework for deploying the mass-production revolution: the centralized, hierarchical pyramid with functional compartments. This structure was applied in the economy by almost every corporation, but was also replicated in any other organization confronted with a large and complex task in government, in hospitals, in universities, in trade unions and political parties, in the West and in the Soviet system, and in developed and developing countries. (Perez 2002: 19)

Perez's argument is one in which the ongoing transformation of capitalism is spurred by technology (creating resources), which combines with a changed 'common sense' (new rules) to produce new structures (rule-resource sets), including both institutions and material goods. Following this argument, the principles of creating a prototypical business in the new logic of the network society are based on the 'common sense' of the computer engineers who created the technological revolution of the microchip. If that is so, the roots of the search engine culture are to be found in Silicon Valley.

Castells (2001) suggests that 'Internet culture', by which he essentially means the culture of Internet developers, based in Silicon Valley, has four aspects. The first aspect is a culture of techno-meritocracy, derived from academic norms, in which discovery, especially the kind of discovery that leads to technical improvement, is the highest value. The second aspect is the 'hacker' culture that values freedom, especially freedom to create, regardless of organisational rules. The third aspect is the culture of the virtual communitarian, which values lack of hierarchy, free communications and self-directed networking as the ideal community structure. The final aspect is the culture of the entrepreneurs. One might call these levels those of the programmer, the artist, the citizen and the businessperson. According to Castells, the entrepreneurial culture is essentially intertwined with the other elements of the Internet culture: 'The strategy is to change the world through technology and to be rewarded with money and power, via the workings of the financial markets' (Castells 2001: 57). He describes a social system in which money fuels technology, and comes to define personal success and intellectual achievement: '[T]he fact that the reward is external (money) rather than internal (Puritan ethic of self-improvement by honest work) has considerable consequences for the culture . . . so that ideas, work, and the personal accumulation of wealth tend to be associated in the same movement' (Castells 2001: 58). The system of entrepreneurship and innovation includes inventors, technologists, and venture capitalists, who 'come together in a process of production and innovation that ultimately creates companies, makes money, and, as a by-product, delivers technology, goods, and services. In this process, the relationship between capital and innovation is internalised'.

Castells describes of the equation in the Silicon Valley culture of technological development and economic success with virtue, and an intense relationship between capital and innovation in the culture of Internet production. This description closely echoes the results of the analysis in the study of search engine producer discourse, where technical prowess, financial success and the persona of the guardian of the web were intertwined in search producer discourse (Van Couvering 2007). The Internet culture Castells describes leaves space for freedom and individual networking but places a premium on individual intellectual achievements. In terms of search engine bias, this same culture holds together the idea of intellectual, technical achievement to achieve high monetary returns and rewards. The public aspect of Internet culture, exemplified in Castell's virtual communitarian, is one in which the individual is free to associate and network on an *ad hoc* basis, not one in which permanent and binding associations are formed, and not one in which public discussion and regulation should place limits on behaviour.

Applying Castell's notion of Internet culture to Perez's framework of 'common sense' economics yields important insights into the rapid development of the search industry: If the new rules which liberate new resources in the network society are those of individual technical prowess, and loose affiliations, we can begin to see the way in which the logic of search classes *all* Internet content as potential nodes in a network and as potential revenue sources. While the old logic of media is built on the industrial model focusing on economies of scope and scale, the new logic suggests a loose affiliation in which clever technology provides competitive advantage. Yet the idea of 'media', defined as a cultural product funded by advertising, was still available. Drawing on these very specific and local stocks of knowledge appears to have enabled the entrepreneurs at Yahoo!, Google and Microsoft to quickly create a new kind of media business, when overlain onto the previous common-sense idea of media. But the new thinking saw other content providers as network resources, rather than as competitors.

In addition to guiding and realising resources in local structuration processes the 'common sense' of the new network paradigm restructures geography in a particular way, relevant to the rapid creation of the global search engine industry. According to Perez, each technological revolution and associated developmental surge begins in a core country and spreads outward, with the outward spread lasting much longer than the whole wave in peripheral countries. Thus, for example, some areas today are still moving to mass production. Perez suggests that the current surge is much more global in character than previous surges, since the common sense of the network paradigm is one which does not to recognise geographic boundaries (this is the 'space of flows' to which Castells refers). In particular, she notes that the spread of trade and productive activities related to the current surge began much earlier than in previous surges, and is, in fact, a feature of this techno-economic paradigm (Perez 2002: 66f).

The US search engines, operating from the Silicon Valley core, began very quickly to offer services in many other countries, although by no means equally in all (Van Couvering 2008a). In this sense search engines are acting very much as a kind of business infrastructure to bring customers, often across borders, to retailers often based in the core countries. In this way, search engines not only expand themselves but also enable the expansion of Internet commerce and culture, particularly that of the western core, which in turn funds their expansion through advertising.

This section has suggested that both the practices of search engine producers and the spread of search engines across the globe are related to the logic of the network, a specific cultural form which developed in and around Silicon Valley. The power of this logic to create new resources and new productive capacity was dramatic, and its strategic use had great benefits to search engine producers, who were able to cast technological issues as issues of economic success or failure. But no cultural form exists on its own: In the complex of the Silicon Valley culture, it is difficult to draw attention to issues, such as search engine bias, which need to be solved through public intervention rather than through market forces or technical innovation.

CONCLUSIONS

Throughout my work on search engine production, certain conflicts have suggested themselves as important to the search engine results. For example, there is the evident conflict between the search engine producers and the search engine optimisers, with certain producers characterising it as a 'war' (Van Couvering 2007). There is the struggle of smaller websites to be represented and highly placed in the results. There is the issue of free provision of information versus restricted access to certain types of content. There are national interests versus international technological provision.

Important though these conflicts are, they are underlain by a fundamental agreement between search engine producers, search engine marketers, search engine advertisers and owners and managers of websites both large and small, full of content both legal and illegal, in many different languages. For these groups, search engines are a business, and search engine results are first and foremost a commodity. This news may come as something of a surprise to the ordinary user who views search engines as a neutral window on the virtual world, but there is no mistaking the evidence. The in my view fundamental conflict between the search engine as a private media enterprise and the search engine as a public good is very much understated, while the resources that the private enterprise brings to bear far outweigh the resources of the ordinary Internet user.

Using discourse analysis as a method of inquiry to complement historical and economic investigation of search engines has led to important insights about the interpenetration of larger systems into everyday scenarios where technology is constructed and interpreted. First, it is apparent that technological schemas, which are the way in which actors account for and give explanations of their actions with reference to technological development and use, are part of larger cultural schemas. These technological schemas and their associated practices are in turn clustered into integrated wholes, each one with its own potential for liberating and allocating resources. Actors have a multiplicity of potentially conflicting schemas and practices from which to draw, and they use particular schemas strategically in order to mobilise different resources.

The mobilisation of resources and therefore the stability or flux of smaller level structures or larger level systems is not therefore a question of economic determinism, but rather of how culture and resources come to constitute each other in different patterns. To date, the political economy of communications has focused primarily on how the economic and material bases of the 'cultural industries' affect culture more widely. Garnham, for example, writes about the analysis of meaning in the service of systems of domination:

> A delimited social group, pursuing economic or political ends, determines which meanings circulate and which do not, which stories are told about what, which arguments are given prominence and what cultural resources are made available and to whom. *The analysis of this process is vital to an understanding of the power relationships involved in culture and their relationship to wider structures of domination.* (cited in Calabrese 2004: 2; emphasis added)

This research shifts the emphasis onto the importance of cultural forms in constructing and maintaining both local structures and wider systems. In other words, the circulation of meaning has as much influence in the culture of production as in the culture of consumption, where much work has been focused. In the case of search engines, the cultural forms of the technological entrepreneur have been extremely effective in creating new mega-businesses out of ideas and abstract thoughts, but much less successful when it comes to ensuring or debating issues of public good or public responsibility on information issues.

This chapter has used search engines as a case, suggesting one method by which the local practices of technology production could be related to larger systems by means of a cultural understanding of structuration theory. The analysis in this chapter shows a complex dynamic at work between systemic factors and agents' actions. It shows that just as insights from the theory of political economy might imply, much of the bias of search is indeed tied in with the search engine company as a business environment and the linkage of technological innovation to business concerns, such as market size and revenue generation. But, crucially, this linkage is not primarily imposed from outside by owners or managers. Rather, it has to do with the creation of meaning on an everyday level within a specific cultural framework. Search engine producers are *participants* in the culture of the Internet, and as such they argue for and create products which in turn reinforce that culture. In one sense, the insight of the some technology scholars that the key to technological change are rhetorical actions (e.g., Bijker 1989) seems validated by the preceding analysis; however, what this chapter shows is that those rhetorical actions are important precisely because they allow resources such as money, computers, and staff to be mobilised to support certain types of technical development. The discussion of rhetoric without resources misses the point, and vice versa. The key to understanding both change and stasis in technological development is to examine the two elements together.

From this perspective, therefore, questions of bias in search are indeed to do with economic concerns and the overall large-scale features of the search engine business, but these concerns do not appear to be (or not wholly) imposed by management. Rather they seem to be naturalised through specific technological schemas that are created through participation in the Internet culture and reified in search results.

NOTES

1. These could be called *relevant social groups* (Bijker et al. 1989).
2. Indeed, Geertz's (1973) pioneering use of 'thick description' puts culture as the object of empirical study squarely in the realm of observable practice.
3. 'Long waves' were first proposed by the Soviet economist Kondratiev in 1925.

REFERENCES

Bergman, M. K. (2001). White Paper: The Deep Web: Surfacing Hidden Value. *Journal of Electronic Publishing*, 7(1).

Bijker, W. E. 1989. The Social Construction of Bakelite: Toward a Theory of Invention. In *The Social Construction of Technological Systems: New Directions in the Sociology and History of Technology*, edited by W. E. Bijker, T. P. Hughes and T. Pinch. Cambridge, MA: MIT Press, 159–187

Bijker, W. E. 1995. *Of Bicycles, Bakelites, and Bulbs: Toward a Theory of Sociotechnical Change*. Cambridge, MA: MIT Press.

Bijker, W. E., T. P. Hughes and T. Pinch. Eds. 1989. *The Social Construction of Technological Systems: New Directions in the Sociology and History of Technology*. Cambridge, MA: MIT Press.

Calabrese, A. 2004. Toward a Political Economy of Culture. In *Toward a Political Economy of Culture: Capitalism and Communication in the Twenty-first Century*, edited by A. Calabrese and C. Sparks. Lanham, MD: Rowman & Littlefield, 1–12.

Castells, M. 2001. *The Internet Galaxy: Reflections on the Internet, Business, and Society*. Oxford: Oxford University Press.

Diaz, A. 2008. Through the Google Goggles: Sociopolitical Bias in Search Engine Design. In *Web Search: Multidisciplinary Perspectives*, edited by A. Spink and M. Zimmer. Berlin: Springer, 11–34.

Geertz, C. 1973. *The Interpretation of Cultures: Selected Essays*. New York: Basic Books.

Giddens, A. 1984. *The Constitution of Society: Outline of the Theory of Structuration*. Cambridge: Polity Press.

Goldman, E. 2006. Search Engine Bias and the Demise of Search Engine Utopianism. *Yale Journal of Law & Technology*, 9, 111–123.

Hargittai, E. 2000. Open Portals or Closed Gates? Channeling Content on the World Wide Web. *Poetics*, 27(4), 233–253.

Introna, L. D. and H. Nissenbaum 2000. Shaping the Web: Why the Politics of Search Engines Matters. *The Information Society*, 16(1), 169–185.

Kleinberg, J. M. 1998. Authoritative Sources in a Hyperlinked Environment. *Ninth Annual ACM-SIAM Symposium on Discrete Algorithms*. Retrieved from http://www.cs.cornell.edu/kleinber/auth.dpf. Last accessed 2003–08–28.

Kleinberg, J. M. and S. Lawrence 2001. The Structure of the Web. *Science*, 294, 1849–1850.

Lawrence, S. and C. L. Giles. 1999. Accessibility of Information on the World Wide Web. *Nature*, 400, 107–109.

Machill, M., C. Neuberger and F. Schindler. 2003. Transparency on the Net: Functions and Deficiencies of Internet Search Engines. *Info: The Journal of Policy, Regulation and Strategy for Telecommunications*, 5(1), 52–74.

Mowshowitz, A. and A. Kawaguchi. 2002. Assessing Bias in Search Engines. *Information Processing and Management*, 38, 141–156.

Mowshowitz, A. and A. Kawaguchi. 2005. Measuring Search Engine Bias. *Information Processing & Management*, 41(5), 1193–1205.

Orlikowski, W. J. 1992. The Duality of Technology: Rethinking the Concept of Technology in Organizations. *Organization Science*, 3(3), 398–427.

Orlikowski, W. J. 2000. Using Technology and Constituting Structures: A Practice Lens for Studying Technology in Organizations. *Organization Science*, 11(4), 404–428.

Orlikowski, W. J. and D. C. Gash 1994. Technological Frames: Making Sense of Information Technology in Organizations. *ACM Transactions on Information Systems*, 12(2), 174–207.

Perez, C. 2002. *Technical Revolutions and Financial Capital: The Dynamics of Bubbles and Golden Ages*. Cheltenham, UK: Edward Elgar.

Potter, J. and M. Wetherell 1987. *Discourse and Social Psychology: Beyond Attitudes and Behaviour*. London: Sage.

Rogers, R. 2000. Introduction: Towards the Practice of Web Epistemology. In *Preferred Placement: Knowledge Politics on the Web*, edited by R. Rogers. Maastricht: Jan van Eyck Akademie, 11–24.

Sewell, W. H. 1992. A Theory of Structure: Duality, Agency and Transformation. *American Journal of Sociology*, 98(1), 1–29.

Van Couvering, E. 2007. Is Relevance Relevant? Market, Science, and War: Discourses of Search Engine Quality. *Journal of Computer Mediated Communication*, 12(3), 866–887.

Van Couvering, E. 2008a. *Finding the Centre: Search Engines in Japan, China, Germany and South Africa*. Paper presented at the International Association of Media & Communications Researchers (IAMCR) 26th Annual Research Conference, Stockholm, Sweden.

Van Couvering, E. 2008b. The History of the Internet Search Engine: Navigational Media and the Traffic Commodity. In *Web Search: Multidisciplinary Perspectives*, edited by A. Spink and M. Zimmer. Berlin: Springer, 177–206.

Vaughan, L. 2004. New Measurements for Search Engine Evaluation Proposed and Tested. *Information Processing & Management*, 40(4), 677–691.

Vaughan, L. and M. Thelwall. 2004. Search Engine Coverage Bias: Evidence and Possible Causes, *Information Processing & Management*, 40(4), 693–707.

9 Technology and Epistemology

Information Policy and Desire

Sandra Braman

The science of science and technology policy is a relatively new funding area for governments. The research agenda is multiply reflexive, asking, 'What knowledge can we produce about the decisions we make about how to produce knowledge?' Directing public funds to support such work is a matter of information policy—laws, regulations and doctrinal positions dealing with information creation, processing, flows and use. Funding programs in this area mark the horizon of epistemological desire for the state, a desire that is Lacanian in nature, enduring and unquenchable.

Since the late 17th century, that desire has been channelled through facticity, a cultural formation oriented around the fact, whether towards or away. Where facticity is important, there are specific social functions for narrative forms that present themselves as factual. Societies dependent upon facticity generate and adhere to detailed and verifiable procedures for the development, presentation, evaluation and use of facts. Institutional authorities certify specific types of information as factual and information production processes as fact producing.

Facticity became dominant as a cultural formation after John Locke introduced the concept of the fact in 1690. For Locke, facts appear when a *perceptual entity* has an *experience* of the material or social environment, symbolically *expresses* what has been learned about the environment, and those referential expressions become the subject of *discussions* through which agreement is reached on what will collectively be accepted as the truth.

The impact of Locke's conceptualisation of the fact was enormous and immediate. His *Essay on Human Understanding* was the most influential book of the 18th century aside from the Bible (MacLean 1936), and the work is considered to have established the foundations of modern epistemology (Foley 1999). Lockean fact catalysed the resolution of distinct genres such as novels, journalism and history out of the previously undifferentiated matrix of narrative forms (Davis 1983). Tom Wolfe (1973) describes the impact as equivalent to that of the introduction of electricity—a parallel that is far from trivial, for the early Industrial Revolution influenced cultural receptivity to facticity by providing evidence of its concrete rewards.

Facticity has thus long been intertwined with technological innovation. For Canadian communication theorist George Grant, technology *is* epistemology, not just extending human abilities but offering a new approach to knowing and making in which both activities are changed (Barney 2000). Economists who rely upon evolutionary theories of innovation also see technology as epistemological in nature since technological change affects what it is that we know, and new inventions—the technological equivalent of biological speciation—allow us to know new things (Mokyr 1990).[1]

From this perspective, information policy can be understood as epistemology policy. This chapter provides a first pass at examining the impact of recent technological innovations on epistemology policy by working through each element of Lockean fact as a distinct policy order that builds upon that which is prior, adding new degrees of freedom in an ourobouros in which the fourth order (discussion) returns us to the first (perception) in today's technological environment. The chapter opens by briefly examining relationships between facticity and the law and information policy as epistemology policy before turning to the specific policy issues that arise as a result of technological innovation at each order.

FACTICITY AND THE LAW

Lennard Davis provides a fascinating account of the role of the law in creating facticity as a social formation with political valence beginning in the 17th century. This period was marked by the tensions that accompanied the shift from religious to secular political power, the politicisation of narrative, and seemingly endless civil wars. He notes,

> The ideologizing of language . . . created the conditions for legal intervention into the realm of the discourse to diffuse the politicizing of news/novels, which then created the conditions for a definition of fact and fiction in which the former could be repressed and the latter more or less ignored. (Davis 1983: 83)

Davis discusses the ways in which specific legal tools shaped the nature of facticity, including libel law; stamp and tax laws; the criminalisation of the production of certain types of narrative and of the lifestyles associated with narrative production; and the licensing, registration and regulation of the means of production of mass communication of the era—printing presses.

Other types of law were essential to facticity during that period as well. The earliest materials to be copyrighted, for example, were factual in nature—maps, calendars, law books and arithmetic and grammar primers (Ginsburg 1990). Over time, the range of laws and regulations that affect the nature of facticity has expanded; facticity is often involved when regulatory agencies move into previously unregulated domains of activity as when the US Federal Trade Commission (FTC) turned its attention to consumer fraud (Biegel 2001).

One of the functions of the law is to 'fix' facts. For a long time, legal procedures and rules of evidence have been among the certified procedures for the development and use of facts. More recently—at least in the US—courts have become judges of quality of scientific evidence, as the extensive literature on the effects of the US Supreme Court decision in *Daubert* details. The law provides authoritative meaning in the face of ambiguity (Catlaw 2006), 'rationality rituals' for the evaluation of risk (Mercer 2008), and decisions as to when scientific controversies have been resolved (Ingram-Waters 2009). Patents fix the boundaries of otherwise mutable objects (Marques 2005). Regulatory agencies institutionalise the mathematical ratios to be used for accounting purposes (Hopwood and Miller 1994). The law affects facticity as a cultural formation even when the subjects of legally mandated information collection and distribution are inherently unknowable (as can be the case for the intelligence establishment (Berkowitz and Goodman 2000). And it can establish uncertainty itself as a legally acceptable foundation for decision-making (as has happened with stock offerings in the US [Fortun 2001]).

Insights from many different disciplines help us understand just how legal interventions affect the nature of facticity and, thus, epistemology. Innis's research on the economic effects of communication media (1951), Leith's (1983) analysis of the history of the standardisation of the English language, Sigal's (1973) history of journalism in the US, Venclova's (1983) review of the history of Soviet government constraints on the production of texts, Power and Laughlin's (1996) use of Habermas's theory of law to investigate the nature of facticity in accounting, and Lash's (2007) analysis of relationships between particular configurations of facticity and political hegemony all make the historical point as manifested in quite diverse circumstances. MacKenzie's (2007) analysis of the interactions between legal and technological innovations that generated confidence in the claimed facticity of derivatives demonstrates the value of this perspective for understanding the 21st century financial crisis. Building on the work of John Dewey, political scientists link the knowledge dimensions of the law to the epistemology of democracy itself (E. Anderson 2007; Ezrahi 1995). Changes in policy-making processes, rules of evidence, constraints on the use of research results as inputs into policy-making, and shifts in the location of the burden of proof are all manipulations of the epistemological conditions of legal decision-making (Gysen et al. 2006).

Epistemological consequences of the effects of technological innovations on the nature of facticity run in both directions. As Nowotny—a historian of the sociology of knowledge—argues, the development of contemporary social science practices between the two 20th century world wars was driven by the desire to influence policy-making and a sense of the present as 'a strategic place from which to influence the future' (Nowotny 1983:

189). In one among many documented concrete examples, the facticity of data about teen automobile accidents and fatalities has influenced social policy, education and governmental responses to risk (Best 2008). Matoesian (1995) argues that policy analysis itself should incorporate attention to interactions between perceptions of facticity and the social construction of illegal behaviour when considering legal reforms.

Legal struggles to deal with the effects of technological change in particular have epistemological consequences since technology is itself a way of knowing or, in Turkle's (2004) words, 'a carrier of a way of knowing'—the technologies we use alter what we know, and what we know about, as well as our relationships with the material and social worlds in which we live. As Heidegger (1977) noted, all innovations have epistemological consequences because each demands that we think in different ways; Postman applied that insight to one specific cultural technology when he commented that 'television is the command centre of the new epistemology' (Postman 1985: 78).

There are, of course, differences in relationships between knowledge and technology across societies (Kosempel 2007) and across individuals (Lee 2008). Thus epistemological consequences of take-up of particular technologies are susceptible to a number of intervening variables. (Unfortunately, 'android epistemology'—the epistemology of machines as they think, as distinct from the impact on human epistemology of machines [Ford, et al., 1995]—is beyond the scope of this chapter.)

Certainly much of the development of information policy since the early 20th century responded to technological change (Braman 2004).[2] By the 1950s, knowledge growth itself had come came to be seen as a contingent process requiring explicit decision-making—including laws, regulations and policies—at every point (Harwood 1993). Since then, governments have directed the knowledge growth that leads to technological change through a number of legal mechanisms (Braman 2007).

With causality running in every direction, it is most accurate to describe relationships among epistemology, technology and the law as co-constructed. From this position, we can explore developments in information policy as epistemology policy driven by the effects of technological change on the cultural formation of facticity.

INFORMATION POLICY AS EPISTEMOLOGY POLICY

The salience of legal interventions into facticity rose with the shift from an industrial to an information society. Just as the economics of information has arrived as a distinct field that brings under one roof diverse theories, concepts and research traditions historically treated as unrelated to each other, so information policy has come to serve as an umbrella term for all laws, regulations and doctrinal positions (constitutional principles) affecting information, communication and culture. Increasingly, even individuals long accustomed to thinking in terms of media or communication policy are turning to the concept of information policy as the phrasing that best captures the range of the legal domain affecting all aspects of media technologies, institutions, content and practices.

Conceiving of information policy as epistemology policy focuses our attention on the impact of legal decision-making on how we know what we know. From the perspective of the study of communication, this is a natural move. Epistemological assumptions are foundational to communication theory (J. Anderson 1996; Krippendorff 1984) and research (Grimes and Bergen 2008; Slack and Allor 1983), underlie specific research questions and methods (Ishii 2004; Pfau 2008), and allow us to see certain media as 'epistemic infrastructure' (Hedrom and King 2005).

It is also a natural move for policy analysis, for epistemic functions are critical to the development and sustenance of governance systems (Adler and Bernstein 2005). Much of the influence of British legal history came through the work of Blackstone, who was himself much influenced by Locke. Benoliel (2004) describes the entire problem of regulating cyberspace as epistemological. Theorists of freedom of expression routinely point to Locke as a foundational thinker (see, e.g., Emerson 1970; Rawls 1971; Schauer 1984), and Locke has been read as an early proponent of information policy (Duff 2004). Scholars are beginning to explicitly analyse issues such as access to information from an epistemological perspective (see, e.g., Des Bordes and Ferdi 2008).

Most discussions of relationships between epistemology and the law, however, treat epistemology as if it were a singular process. From a social science perspective, however, there are at minimum four separable epistemological moments, or orders, as distinguished by the four elements of Lockean fact. Factors that prompt shifts in our epistemological theories and practices, or in the consequences of having taken particular epistemological positions, may come into play at only one of these moments. The enormously important political differences between state 'knowledge' of a hotly contested election—say, that of El Salvador in 1982—and what an individual reporter 'knows' of the same event, for example, are born quite distinctly at the moment of the definition of the perceptual entity involved (Braman 1984, 1985). First-order epistemology policy issues affect the nature of the perceptual entity. Second-order issues affect the way in which the material and social worlds are experienced. Third-order issues affect expression, the translation of experience into facts. And fourth-order issues affect the way in which facts produced by individual perceptual entities are discussed within social groups, ultimately yielding a consensually shared understanding of the truth—what Berger and Luckmann (1966) so influentially described as the social construction of reality.

Full analysis of the four orders of epistemology policy as affected by technological innovations would include a look at the Lockean foundation, pertinent communication theories, what is known about the impact of technological innovation, the history of thinking about information policy issues from an epistemological perspective (where such exists), and a reframing of contemporary information policy issues from an epistemological perspective. Here, there is space for only a brief introduction to each.

FIRST-ORDER EPISTEMOLOGY POLICY: THE PERCEPTUAL ENTITY

Epistemological desire begins with longing to know, to be consciously and self-consciously aware of information and to be able to hold it to oneself or to share. For Locke, the perceptual entity did not necessarily have to be an individual—it need only be a consciousness which endures across time:

> Self is that conscious thinking thing (whatever substance made up of, whether spiritual or material, simple or compounded, it matters not) which is sensible, or conscious of pleasure and pain, capable of happiness or misery, and so is concerned for itself, as far as that consciousness extends. (Locke 1924: 194)

A Lockean individual is a locus of consciousness which may reside in a single human being, a group of human beings, or have no physical manifestation at all. Corporations, geopolitical governments, communities and social networks can all be loci of consciousness for epistemological purposes.

Clear identification of the entity responsible for any given perception remains key to facticity, whether that is in the course of journalistic (Vultee 2010) or political (Rapley 1998) practice. However, until the mid-19th century, the locus of consciousness was almost universally understood to reside in the individual in Western thought. With the growth of the bureaucratic welfare state and the professionalisation of scholarship and research, from the mid-19th century through the mid-20th century, the critical locus of consciousness then came to be commonly placed in public authorities. The tension between these two perspectives set the stage for the battles between 'new' and 'objective' journalism of the 1960s and 1970s (Braman 1984, 1985). In retrospect, these disputes over the production of facts by journalists were early empirical indicators of the fundamental shifts in the nature of facticity that have been the subject of so much post-modern theory. By the last decades of the 20th century, groups—whether organisations, communities or networks—became an alternative to individuals and the state as perceptual entities that were understood to be the sources of facts. Theories of 'social epistemology' take the Lockean notion of discussion as critical to discerning the truth by exploring the epistemic interdependence of individuals and the resulting epistemic division of labor (Fuller 2002), and the nature of epistemic agency (Damsa et al. 2010; Eyal and Buchholz 2010).

The transition to an information society has affected the nature of the Lockean loci of consciousness in several ways. Technologies have facilitated the development of forms of social organisation that present new types of 'knowers', such as firms that are informational mechanisms (Coase 1937) with cognitive capacity (Eisenberg 1984), epistemic networks (Leydesdorff 1991) and epistemic communities (Hakanson 2010). Research on the current version of the 'convergence of technologies'—the convergence of nanotechnology, biotechnology, information technology and cognitive technology, or NBIC—has drawn attention to possibilities for technological interventions into the cognitive capacities of humans, affecting the nature of humans, whether individually or in groups, as perceptual entities (Canton 2004; Roco 2004).

The reliance of such technologically mediated Lockean loci of consciousness on technical details means that the law-like nature and effects of technological design and network architecture (e.g., Biegel 2004; Braman 2010) are quite critical from the perspective of epistemology policy. It is these details that provide the affordances for the social networks that have been understood as important to the making and implementation of laws since at least the early 1970s (Moore 1973). Social networks can be functional alternatives to contract law (Landa 1981), they have utility in bringing groups under the rule of law (Hagan 1998) and they can be policy tools used to implement the law (Boykin and Roychowdhury 2005). The law influences social network processes, whether the activities involved are illegal (Baker and Faulkner 1993) or legal (Rauch 2001); antitrust, or competition, law is a particularly important and always problematic example of the types of legal forces that attempt to create boundaries between illegal and legal forms of social networking (Kilduff and Tsai 2003).

The rights of association and assembly are premiere examples of doctrinal principles for information policy with epistemological implications for the nature of the perceptual organism; these principles frequently trigger legal and regulatory problems at the point of interpretation and implementation as they do in the problem of identifying the legal subject in cases involving WikiLeaks (Braman, in press). Intellectual property rights can also affect the nature of the perceiver of facts, as when software through which group interactions take place is patented, or laws are passed that deny Internet access to those accused of copyright infringement before there has been any determination of actual guilt. Legal interventions into various literacies would also fall into this category.

SECOND-ORDER EPISTEMOLOGY POLICY: EXPERIENCE

Epistemological desire may be most evident when it comes to experiencing the world. As legal theorist Unger put it, 'We live among our particulars, but we always want and see something more than any particular can give or reveal' (Unger 2007: 21). It is here that the imagination drives fantasies of transcending the species by achieving 'cognitive liberty'; analysing all scientific data across time, space and discipline simultaneously; or, at minimum, knowing everything there is to know about every citizen.

When Locke was writing, experiences reported upon as facts were understood to be directly acquired through the senses; the systematic practices for doing so that we now call 'science' had not yet been standardised or acquired normative status. The emphasis on experience as how we know the world was part of the shift towards empiricism and away from the Platonic view that perceptions are based on eternal forms (Peters 1989). With sensation as the source of all knowledge, it is the stimuli of the physical world that are of central interest (Laurence 1980). What the locus of consciousness receives via the experience of the senses is the basis of fact, and facts communicated to others are experienced by the senses of others. Locke's ideas inspired other thinkers to attempt to develop distinctions among types of experiences as they generate facts (O'Neal 1996).

Locke did, however, acknowledge limits to what the locus of consciousness can learn from experience: Sensory perceptions do not translate directly into the ideas that are expressed, some ideas appear in our minds without an apparent sensory stimulus, painful sensory perceptions may distort ideas, and what we experience as the evidence of our senses may instead be what we expect to find based on what we have been told by others (Osler 1970). Communication theorists have taken this much further, not only in their sociological, psychological and physiological analyses, but also in their thinking about the impact of technology on the nature of experience. Putting to the side naive views of technology as neutral, at least four different approaches have appeared. Frankfurt School thinkers such as Walter Benjamin (1936/2006) argued that technology is *a barrier to experience*. Heidegger is generally credited with being the first to introduce the view that, instead, technologies *shape experience* (Barney 2000), an approach that has informed both macro-level analyses (e.g., Mumford 1934; Innis 1951) and micro-level analyses (e.g., Brey 2005; Novak 1997). McLuhan's (1964) work epitomises the view of technology *as experience*, a perspective that focuses on the ways in which technological extensions of human senses changes the way in which such Lockean loci of consciousness come to know the empirical world. For post-modern thinkers such as Baudrillard (1983), technologies are *the reality that is experienced*. The trajectories of empirical research about the effects of technologies support theoretical shifts among those perspectives; Floridi (2005), for example, notes that studies of telepresence began with the assumption that it involved epistemic failure, but over time have come to focus instead on ways in which telepresence provides effective means of observation and other experience.

Whichever of these positions one takes theoretically, there is no doubt that digital technologies raise a number of epistemology policy issues deriving from the effects of these technologies on the nature of experience. Some of these involve restrictions: Internet filtering, like all censorship, restricts Lockean experience and thus the facts that result. Technologies can also expand experience through such techniques as the embedding of intelligent networked sensors in the material environment; laws and regulations for ambient embedded computing are still being developed. Privacy law can be understood as establishing limits to the ability to experience other individuals. Government policies regarding the funding of research, and the policy issue with which this chapter began—funding for the production of knowledge about science and technology policy—fall into this category as well.

THIRD-ORDER EPISTEMOLOGY POLICY: EXPRESSION

For Locke, expression was the process through which experience was shared. Though some communication law scholars have distinguished private expression from public communication (Glasser 1991), it is more common to understand his theories about expression as the means by which the private and public were connected in order to be the subject of the discussion that was necessary to discern the truth (Schauer 1984). Indeed, Emerson (1970) understands Locke's ideas about expression to have been developed self-consciously in resonance with early ideas about the nature of science. Locke's ideas about the centrality of expression to facticity were foundational for thinking about freedom of expression in terms of a 'marketplace of ideas', although he himself did not use the metaphor (Schauer 1984).

Communication research deals with epistemological dimensions of expression when it looks at such topics as the effects of representation (Markova 2008), motivations for communication (Knobloch et al. 2003) and contextual dimensions that affect the reception of communication (Ekbia and Maguitman 2001). Research based on medium theory (Meyerowitz 1994) addresses the effects of differences among technologies on both expression and how we make meaning out of the expressions of others (Jones 2006). Examples of such work include comparisons between the broadcast and print news (Silcock and Keith 2006) and analysis of differences between the evidentiary availability of information when it is collected and distributed through diverse technologies (Ettema and Glasser 1998).

There are many examples of epistemology policy issues either generated or exacerbated by technological innovations in the digital realm. Legal issues involving epistemological facets of expression, such as treason, fraud, truth in advertising and libel, often vary significantly—and can be variously problematic—depending on the technology involved (Braman 2004). New types of epistemological issues that have appeared as the result of technological innovation include authentication of the identity of authors and of expressions themselves as digital objects (Jantz 2009). Surveillance practices raise a range of epistemological issues for policy-makers, particularly when the information collected is not about actual expressions or behaviours but, rather, inferences (Strickland 2005). Data retention affects epistemology by transforming expressions meant to be ephemeral and directed at only one or a few for a particular purpose into expressions that endure over time and can be interpreted as if directed at other receivers entirely.

FOURTH-ORDER EPISTEMOLOGY POLICY: DISCUSSION

The fourth element of Lockean fact is discussion of the facts expressed by many people so that the truth can collectively be determined. The face to face interpersonal and group conversations, and extended written exchanges—including those that are published—that Locke would have been thinking about when he emphasised the role of group discussion are still critical to what we are now more likely to call 'knowledge production' than 'truth'; even Habermas claimed that the ideal speech situation would produce epistemological truths (Premfors 1992). Locke actually offered the first in-depth philosophical treatment of the term 'communication' as a central principle of speech and language. His ideas were foundational to the modern notion of communication (Peters 1999)—even if he was over-optimistic about the ability of group discussion to create a public consensus on matters of policy (Held 1989) and about the extent to which public consent rests on government fulfilment of its commitments to citizens (Carter 1985).

Communication researchers come to Lockean expression and facticity through sociology. The origins of analysis of the effects of technological innovation on Lockean discussion may be located in the work of Tarde (Hardt 1979), with his concern about the inability of individuals to meaningfully make sense out of, and use, the increased flow of information that came with electrification. Locke's ideas about epistemology influenced the early 19th century social science origins of communication research (Smith 1991), as well as later thinkers such as Durkheim, who emphasised the social origins of epistemological understanding, who in turn had an impact on significant researchers and theorists of the late 20th century (Douglas 1986; Rawls 1971). His conceptualisation of public opinion (which he described as the 'law of fashion') dominated 20th century discussions of what freedom of expression entails and how it should be operationalised (Noelle-Neumann 1984). Later sociologists sustained the connection, as when Bourdieu (1991) noted that communication technologies, rather than new types of scientific instrumentation, would have the most significant impact on the scientific community because of its influence on the nature of scientific reasoning. The inapplicability of Western communication theories across cultures stems largely from epistemological issues that have sociological foundations.

Since Locke's time, of course, the sheer volume of such conversations has multiplied by many orders of magnitude. Each variant on a digital communication technology brings experimentation with new forms of communication that have epistemological implications. Theory and research have transformed the concept of 'discussion' into a complex forest of driven flows of messages and cynical or purposeful reception and meaning-making shaped by multiple forces. Today, research in this area is rife, ranging from investigations into the growing epistemological burden borne by individuals as a result of the weakening of those institutions that historically certified facticity (Arazy and Kopak 2011) to the replacement of social cooperation with commodity exchange as the dominant epistemic influence (Parker 1994), the epistemological functions of trust—also important to Locke (Dunn 1984)—in technologically mediated communication (Pila 2009), and new opportunities for deceit (Daniels 2009). A number of epistemological problems have been generated by collaborative and team-based research communications (Mauthner and Doucet 2008). There have been shifts in the nature of discussion about knowledge provided by subject experts (Aligica and Herritt 2009), the appearance of genres such as Wikipedia that introduce qualitatively new procedures for epistemological discussion (Tollefson 2009) and intensification of interest in knowledge codification because of its value to community discussion (Steinmueller 2000). Other contemporary communication research on the epistemological dimensions of Lockean expression includes work on the effects on facticity of interpersonal communication and affect (Murakami, 2006), the relative weight of practice and epistemological perceptions in determining the utility of facts as expressed (Marques 2005), the impact of epistemic stance on the effects of knowledge organisation on perceptions of facts (Tennis 2008), political culture and epistemological practice (Federowicz 1988) and the epistemological dimensions of participatory communication (Liao 2006).

Fourth-order epistemology policy issues involving Lockean discussion lead us directly to the social construction of reality. We are accustomed to thinking about rights to association and assembly from a communication law perspective, but through the epistemology policy lens we can go further in identifying legal issues that affect our uses of discussion to figure out, or construct, what it is that we believe we know. The issues are myriad, difficult and diverse. They are raised by such developments as changes in rules of evidence, administrative constraints on the use of research as inputs into policy-making, media concentration and network neutrality—all of which can affect the epistemological qualities of discussion by restricting the range of voices and types of information heard. The relative weight of particular types of voices in public discussion is affected by election campaign regulations dealing with the amount corporations can spend—and whether or not corporations can speak anonymously[3]—during elections, time limits and geographic constraints on campaigning. Laws and regulations pertaining to social media can constrain or prohibit certain types of public discussion either directly or indirectly, through a 'chilling' effect that makes particular types of speech unlikely even if they are not explicitly illegal. Contractual terms of service agreements for participation in social networks, or going online at all, also affect the qualities of Lockean discussion.

One could go on. Doing so leads us back, in an interesting way, for in the digital era information policy—epistemology policy—issues that affect Lockean discussion return us to those that can influence the structure and characteristics of the perceptual organism. We again find ourselves looking at regulatory requirements for social networking software or permitting the patenting of software that provides the affordances for epistemological collaborations.

CONCLUSIONS

This chapter opened with the new area of government funding for knowledge production about how we make decisions about knowledge production—research and development in the science of science and technology policy. It is relatively easy to see this as an example of epistemology policy, but the argument has been made here that many—perhaps all—information policy issues have epistemological dimensions. Treating such issues as epistemological in nature offers the advantage of changing the game, providing a means of shaking up a rigid discourse in which most players chose their seats long ago.

There are matters not only of the sociology of knowledge, but also of politics. Throughout history epistemological crises and turning points accompany political shifts. It is no coincidence that information policy has come to the foreground during this period of profound transformations in law–state–society relations. The stakes are high, and the nature of the moment is clear at the top of the power structure. As former President George W. Bush's strategist Karl Rove said in an oft-republished quote that first appeared in *The New York Times Magazine*,[4]

> [G]uys like me were 'in what we call the reality-based community,' which he defined as people who 'believe that solutions emerge from your judicious study of discernible reality' . . . 'That's not the way the world works anymore,' he continued. 'We're an empire now, and when we act, we create our own reality. And while you're studying that reality—judiciously, as you will—we'll act again, creating other new realities, which you can study too, and that's how things will sort out. We're history's actors . . . and you, all of you, will be left just to study what we do. (Suskind 2004)

One of the tricks of desire is that we may be aware that we crave, but not genuinely understand just what it is that we are longing for. If information policy is the expression of epistemological desire by the state, then those who do information policy analysis must remember to attend to the object of the lust.

NOTES

1. Let's not overlook the bookshelf as a cultural technology with profound epistemological implications. For decades, university bookstores like that at the University of Minnesota presented course readings as offprints, organised class by class. One could wander the shelves and browse what was being read during any given semester across disciplines and classes. Having had the practice of doing this over and over, semester after semester, it was in this way that I encountered Clifford Geertz's (1983) 'Fact and Law in Comparative Perspective', the piece that first informed my understanding of facticity. I have no memory at all of the class for which that piece had been assigned—it wasn't one that I was taking. Similarly, I first encountered Foucault's (1972) *Archaeology of Knowledge* when simply browsing the university's library bookshelves for works on the history of objectivity. It is, of course, getting harder and harder to find bookshelves, and if evidence from my own students is any indicator, the ability to browse bookshelves is a technological skill that has largely disappeared.
2. The word 'change', and not 'innovation', is used here because the introduction of a previously existing technology into a new environment, decay or alteration of existing technologies, or simply swapping out one long-available item for another can also affect the nature of facticity.
3. The 2010 US Supreme Court case of *Citizens United v Federal Election Commission* dramatically altered the election landscape in the US by removing limits on corporate election spending and permitting them to finance campaign advertising and activities anonymously.
4. In the Suskind article, the quote is attributed to an 'unnamed aide'. However, Danner (2007) reported that it was widely known in Washington circles that the source of the quote was Rove.

REFERENCES

Adler, E. and S. Bernstein. 2005. Knowledge in Power: The Epistemic Construction of Global Governance. In *Power in Global Governance*, edited by M. Barnett and R. Duvall. Cambridge: Cambridge University Press, 294–317.

Aligica, P. D. and R. Herritt 2009. Epistemology, Social Technology, and Expert Judgment: Olaf Helmer's Contribution to Future Research. *Futures*, 41, 253–259.

Anderson, E. 2006. The Epistemology of Democracy. *Episteme*, 3(1), 8–22.

Anderson, J. 1996. The Nature of the Individual. In *Communication theory: Epistemological foundations*. New York: Guilford Press, 78–101.

Arazy, O. and R. Kopak. 2011. On the Measurability of Information Quality, *Journal of the American Society for Information Science and Technology*, 62(1), 89–99.

Baker, W. E. and R. R. Faulkner. 1993. The Social Organization of Conspiracy: Illegal Networks in the Heavy Electrical Equipment Industry, *American Sociological Review*, 58(6), 837–860.

Barney, D. 2000. *Prometheus Wired: The Hope for Democracy in the Age of Network Technology*. Chicago: University of Chicago Press.

Baudrillard, J. 1983. *Simulations*. New York: Semiotext(e).

Benjamin, W. 1936/2006. The Work of Art in the Age of Mechanical Reproduction. In *Visual Culture: Experiences in Visual Culture*, edited by J. Morra. New York: Routledge, 114–137.

Benoliel, D. 2004. Technological Standards, Inc.: Rethinking Cyberspace Regulatory Epistemology, *California Law Review*, 92, 1069–1116.

Berger, P. L. and T. Luckmann. 1966. *The Social Construction of Reality*. Garden City, NY: Doubleday.

Berkowitz, B. D. and A. E. Goodman. 2000. *Best Truth*. New Haven, CT: Yale University Press.

Best, A. L. 2008. Teen Driving as Public Drama. *Journal of Youth Studies*, 11(6), 651–669.

Biegel, S. 2001. *Beyond Our Control?* Cambridge, MA: MIT Press.

Bourdieu, P. 1991. *Language and Symbolic Power*. Translated by G. Raymond & M. Adamson. Cambridge, MA: Harvard University Press.

Boykin, P. O. and V. P. Roychowdhury. 2005. Leveraging Social Networks to Fight Spam. *Computer*, 38(4), 61–68.

Braman, S. 1984. *The Location of the Lockean Consciousness in News Reports from El Salvador*. Unpublished MA thesis, University of Minnesota.

Braman, S. 1985. The 'Facts' of El Salvador According to Objective and New Journalism. *Journal of Communication Inquiry*, 13(2), 75–96.

Braman, S. 2004. Where has Media Policy Gone? *Communication Law & Policy*, 9(2), 153–182.

Braman, S. 2007. *Change of State: Information, Policy, and Power*. Cambridge, MA: MIT Press.

Braman, S. 2010. The Interpenetration of Technical and Legal Decision-Making for the Internet. *Information, Communication, & Society*, 13(3), 309–324.

Braman, S. In press. Revelations: Information Policy, the legal subject, and the WikiLeaks Complex. In The Wikileaks Reader, edited by C. Christensen. Frankfurt am Main: Peter Lang.

Brey, P. 2005. The Epistemology and Ontology of Human–Computer Interaction. *Minds and Machines*, 15, 383–398.

Carter, L. H. 1985. *Contemporary Constitutional Lawmaking: The Supreme Court and the Art of Politics*. New York: Pergamon Press.

Canton, J. 2004. Designing the Future: NBIC Technologies and Human Performance Enhancement. *Annual of the New York Academy of Sciences*, 1013, 186–198.

Catlaw, T. J. 2006. Authority, Representation, and the Contradictions of Post-traditional Governing. *American Review of Public Administration*, 36(3), 261–287.

Citizens United v. Federal Election Commission, 130 S.Ct. 876 (2010).

Coase, R. 1937. The Nature of the Firm. *Economica*, 4, 386–405.

Daniels, J. 2009. Cloaked Websites: Propaganda, Cyber-Racism and Epistemology in the Digital Era. *New Media & Society*, 11(5), 658–683.

Damsa, C. I., P. A. Kirschner, J .E. B. Andriessen, G. Erkens and P. H. M. Sins. 2010. Shared Epistemic Agency: An Empirical Study of an Emergent Construct. *Journal of the Learning Sciences*, 19(2), 143–186.

Danner, M. 2007. Words in a Time of War: On Rhetoric, Truth, and Power. In *What Orwell Didn't Know: Propaganda and the New Face of American Politics*, edited by A. Szántó. New York: Public Affairs, 16–36.

Davis, L. J. 1983. *Factual Fictions: The Origins of the English Novel*. New York: Columbia University Press.

Des Bordes, A. and S. Ferdi. 2008. Do Knowledge and New Technologies need a New Epistemology? Presented to the BOBCATSSS Symposium, Berlin.

Available at http://edoc.hu-berlin.de/conferences/bobcatsss2008/. Last accessed 2011-10-05.

Douglas, M. 1986. *How Institutions Think*. Syracuse, NY: Syracuse University Press.

Duff, A. S. 2004. The Past, Present, and Future of Information Policy. *Information, Communication, & Society*, 7(1), 69–87.

Dunn, J. 1984. The Concept of 'Trust' in the Politics of John Locke. In *Philosophy in History: Essays on the Historiography of Philosophy*, edited by R. Rorty, J. B. Schneewind and Q. Skinner. Cambridge: Cambridge University Press, 279–302.

Eisenberg, E. 1994. Dialogue as Democratic Discourse: Affirming Harrison. In *Communication Yearbook*. Vol. 17, edited by S. Deetz. Newbury Park, CA: Sage, 275–284.

Ekbia, H. R. and A. G. Maguitman. 2001. Context and Relevance: A Pragmatic Approach. In *Modeling and Using Context*, edited by V. Akman, P. Bouquet, R. Thomason and R. A. Young. Berlin: Springer, 156–169.

Emerson, T. I. 1970. *The System of Freedom of Expression*. New York: Random House.

Ettema, J. S. and T. L. Glasser. 1998. *Custodians of Conscience: Investigative Journalism and Public Virtue*. New York: Columbia University Press.

Eyal, G. and L. Buchholz. 2010. From the Sociology of Intellectuals to the Sociology of Interventions. *Annual Review of Sociology*, 36, 117–137.

Ezrahi, Y. 1995. Technology and the Civil Epistemology of Democracy. In *Technology and the Politics of Knowledge*, edited by A. Feenberg and A. Hannay. Bloomington: Indiana University Press, 159–174.

Eyal, G. and L. Buchholz. 2010. From the Sociology of Intellectuals to the Sociology of Interventions. *Annual Review of Sociology*, 36, 117–137.

Ezrahi, Y. 1995. Technology and the Civil Epistemology of Democracy. In *Technology and the Politics of Knowledge*, edited by A. Feenberg and A. Hannay. Bloomington: Indiana University Press, 159–174.

Floridi, L. 2005. The Philosophy of Presence. *Presence*, 14(6), 656–667.

Foley, R. 1999. Locke and the Crisis of Postmodern Epistemology. In *New Directions in Philosophy*, edited by P. A. French and H. K. Wettstein. Malden, MA: Blackwell Publishers, 1–20.

Ford, K. M., C. Glymour and P. J. Hayes. Eds. 1995. *Android Epistemology*. Menlo Park: AAAI Press.

Fortun, M. 2001. Mediated Speculations in the Genomics Futures Markets. *New Genetics and Society*, 20(2), 139–156.

Foucault, M. 1972. *The Archaeology of Knowledge & The Discourse on Language*. Translated by A. M. Sheridan Smith. New York: Pantheon Books.

Fuller, S. 2002. *Social Epistemology* (2nd ed.). Bloomington: Indiana University Press.

Geertz, C. 1983. *Local Knowledge*. New York: Basic Books.

Ginsburg, J. C. 1990. Creation and Commercial Value: Copyright Protection of Works of Information. *Columbia Law Review*, 90, 1865–1938.

Glasser, T. L. 1991. Communication and the Cultivation of Citizenship. *Communication*, 12(4), 235–248.

Grimes, T. and L. Bergen. 2008. The Epistemological Argument Against a Causal Relationship between Media Violence and Psychopathic Behavior Among Psychologically Well Viewers. *American Behavioral Scientist*, 51, 1137–1154.

Gysen, J., H. Bruyninckx and K. Bachus. 2006. Evaluating Effectiveness of Environmental Policies. *Evaluation*, 12(1), 95–118.

Hagan, J. M. 1998. Social Networks, Gender, and Immigrant Incorporation: Resources and Constraints. *American Sociological Review*, 63(1), 55–67.

Hakanson, L. 2010. The Firm as an Epistemic Community: The Knowledge-Based View Revisited. *Industrial and Corporate Change*, 19(6), 1801–1828.
Hardt, H. 1979. *Social Theories of the Press: Early German and American Perspectives*. Beverly Hills, CA: Sage.
Harwood, J. 1993. *Styles of Scientific Thought: The German Genetics Community, 1900–1933*. Chicago: University of Chicago Press.
Hedrom, M. and J. L. King. 2005. *Epistemic Infrastructure in the Rise of the Knowledge Economy*. Paris: OECD.
Heidegger, M. 1977. *Basic Writings*. Translated by D. F. Krell. New York: Harper & Row.
Held, D. 1989. *Political Theory and the Modern State: Essays on State, Power, and Democracy*. Stanford, CA: Stanford University Press.
Hopwood, A. G. and P. Miller. Eds. 1994. *Accounting as Social and Institutional Practice*. Cambridge: Cambridge University Press.
Ingram-Waters, M. C. 2009. Public Fiction as Knowledge Production. *Public Understanding of Science*, 18(3), 292–308.
Innis, H. A. 1951. *The Bias of Communication*. Toronto: The University of Toronto Press.
Ishii, S. 2004. Proposing a Buddhist Consciousness-Only Epistemological Model for Intrapersonal Communication Research. *Journal of Intercultural Communication Research*, 33(1–2), 63–76.
Jantz, R. 2009. An Institutional Framework for Creating Authentic Digital Objects. *International Journal of Digital Curation*, 1(9), 71–83.
Jones, J. P. 2006. A Cultural Approach to the Study of Mediated Citizenship. *Social Semiotics*, 16(2), 365–383.
Kilduff, M. and W. Tsai. 2003. *Social Networks and Organizations*. London: Sage.
Knobloch, S., M. Hastall, D. Zillmannand and C. Callison. 2003. Imagery Effects on the Selective Reading of Internet Newsmagazines. *Communication Research*, 30(1), 3–29.
Kosempel, S. 2007. Interaction Between Knowledge and Technology: A Contribution to the Theory of Development. *Canadian Journal of Economics*, 40(4), 1237–1260.
Krippendorff, K. 1984. An Epistemological Foundation for Communication. *Journal of Communication*, 34: 21–36.
Landa, J. T. 1981. A Theory of the Ethnically Homogeneous Middleman Group: An Institutional Alternative to Contract Law. *Journal of Legal Studies*, 10(2), 349–362.
Lash, S. 2007. Power after Hegemony. *Theory, Culture and Society*, 24(3), 55–78.
Laurence, D. 1980. Jonathan Edwards, John Locke, and the Canon of Experience. *Early American Literature*, 15(2), 107–123.
Lee, C. J. 2008. Applied Cognitive Psychology and the 'Strong Replacement' of Epistemology by Normative Psychology. *Philosophy of the Social Sciences*, 38(1), 55–75.
Leith, D. 1983. *A Social History of English*. London: Routledge & Kegan Paul.
Leydesdorff, L. 1991. In Search of Epistemic Networks. *Social Studies of Science*, 21(1), 71–110.
Liao, H.-A. 2006. Toward an Epistemology of Participatory Communication: A Feminist Perspective. *The Howard Journal of Communications*, 17, 101–118.
Locke, J. 1690/1924. *An Essay Concerning Human Understanding*. Abridged and edited by A. S. Pringle-Pattison. Oxford: Clarendon Press.
MacKenzie, D. 2007. The Material Production of Virtuality: Innovation, Cultural Geography and Facticity in Derivatives Markets. *Economy and Society*, 36(3), 355–376.

MacLean, K. 1936. *John Locke and English Literature of the Eighteenth Century.* New Haven, CT: Yale University Press.

Markova, I. 2008. The Epistemological Significance of the Theory of Social Representations. *Journal for the Theory of Social Behaviour*, 38(4), 461–487.

Marques, I. da C. 2005. Cloning Computers: From Rights of Possession to Rights of Creation. *Science as Culture*, 14(2), 139–160.

Matoesian, G. M. 1995. Language, Law, and Society: Policy Implications of the Kennedy-Smith Rape Trial. *Law & Society Review*, 29(4), 699–701.

Mauthner, N. S. and A. Doucet 2008. 'Knowledge Once Divided Can Be Hard to Put Back Together Again'. *Sociology*, 42, 971–985.

McLuhan, M. 1964. *Understanding Media.* New York: McGraw-Hill.

Mercer, D. 2008. Science, Legitimacy, and 'Folk Epistemology' in Medicine and Law. *Social Epistemology*, 22(4), 405–423.

Meyerowitz, J. 1994. Medium Theory. In *Communication Theory Today*, edited by D. Crowley and D. Mitchell. Cambridge: Polity Press, 50–77.

Mokyr, J. 1990. *The Lever of Riches.* New York: Oxford University Press.

Moore, S. F. 1973. Law and Social Change. *Law & Society Review*, 7(4), 719–746.

Mumford, L. 1934. *Technics and Civilization.* New York: Harcourt, Brace and Company.

Murakami, Y. 2006. Horizons of Affectivity. *Studia Phenomenologica*, 6, 17–30.

Noelle-Neuman, E. 1984. *The Spiral of Silence.* Chicago: University of Chicago Press.

Nowotny, H. 1983. Marienthal and After. *Knowledge*, 5, 169–192.

Novak, M. 1997. Transmitting Architecture. In *Digital Delirium*, edited by A. Kroker and M. Kroker. New York: St. Martin's Press, 260–271.

O'Neal, J. C. 1996. *The Authority of Experience.* State Park: Pennsylvania State University Press.

Osler, M. J. 1970. John Locke and the Changing Ideal of Scientific Knowledge. *Journal of the History of Ideas*, 31(1), 3–16.

Parker, I. 1994. Commodities as Sign-Systems. In *Information and Communication in Economics*, edited by R. E. Babe. Boston: Kluwer Academic Publishers, 69–91.

Peters, J. D. 1989. John Locke, the Individual, and the Origin of Communication. *The Quarterly Journal of Speech*, 75(4), 387–399.

Peters, J. D. 1999. *Speaking into the Air.* Chicago: University of Chicago Press.

Pfau, M. 2008. Epistemological and Disciplinary Intersections. *Journal of Communication*, 58, 597–602.

Pila, J. 2009. Authorship and E-Science. *Social Epistemology*, 23(1), 1–24.

Postman, N. 1985. *Amusing Ourselves to Death.* New York: Viking.

Power, M. and R. Laughlin. 1996. Habermas, Law and Accounting. *Accounting Organizations and Society*, 21(5), 441–465.

Premfors, R. 1992. Knowledge, Power, and Democracy: Lindblom, Critical Theory, and Postmodernism. *Knowledge, Technology, and Policy*, 5(2), 77–93.

Rapley, M. 1998. Just an Ordinary Australian. *British Journal of Social Psychology*, 37, 325–344.

Rauch, J. E. 2001. Business and Social Networks in International Trade. *Journal of Economic Literature*, 39(4), 1177–1203.

Rawls, J. 1971. *A Theory of Justice.* Cambridge, MA: Harvard University Press.

Roco, M. C. 2004. Science and Technology Integration for Increased Human Potential and Societal Outcomes. *Annual of the New York Academy of Sciences*, 1013, 1–16.

Schauer, F. 1984. *Free Speech.* New York: Cambridge University Press.

Sigal, L. V. 1973. *Reporters and Officials.* Lexington, MA: D. C. Heath.

Silcock, B. W. and S. Keith. 2006. Translating the Tower of Babel? *Journalism Studies*, 7(4), 619–618.

Slack, J. D. and M. Allor. 1983. The Political and Epistemological Constituents of Critical Communication Research. *Journal of Communication*, 33(3), 208–218.

Smith, W. D. 1991. *Politics and the Science of Culture in Germany, 1840–1920*. New York: Oxford University Press.

Steinmueller, W. E. 2000. Will New Information and Communication Technologies Improve the 'Codification' of Knowledge? *Industrial and Corporate Change*, 9(2), 361–376.

Strickland, L. S. 2005. *Secure Information Sharing*. Washington, DC: Information Security Oversight Office, United States National Archives and Record Administration.

Suskind, R. 2004. Faith, Certainty and the Presidency of George W. Bush. *The New York Times Magazine*, October 17. Available at www.nytimes.com/2004/10/17/magazine/17BUSH.html. Last accessed 2011-04-25.

Tennis, J. T. 2008. Epistemology, Theory, and Methodology in Knowledge Organization. *Knowledge Organization*, 35(2–3), 102–112.

Tollefson, D. 2009. Wikipedia and the Epistemology of Testimony. *Episteme*, 6(1), 8–24.

Turkle, S. 2004. How Computers Change the Way We Think. *The Chronicle of Higher Education*, January 30. Available at http://chronicle.com/article/How-Computers-Change-the-Way/10192. Last accessed 2011-02-02.

Unger, R. M. 2007. *The Self Awakened*. Cambridge, MA: Harvard University Press.

van Eijck, M. and N.X. Claxton. 2009. Rethinking the Notion of Technology in Education. *Science Education*, 93(2), 218–232.

Venclova, T. 1983. The Game of the Soviet Censor. *New York Review of Books*, March 31, 34–35.

Vultee, F. 2010. Credibility as a Strategic Ritual. *Journal of Mass Media Ethics*, 25(1), 3–18.

Wolfe, T. 1973. *The New Journalism*. New York: Harper & Row.

Part III

Uses of Cultural Technologies

In another paper written at approximately the same time, Foucault (1988) describes the practices of the self as '*technologies* of the self' and begins to trace the genealogy of these technologies throughout the history of Western civilisation up to early Christianity. Here he offers very concrete examples and descriptions of the forms that care for the self has taken in different historical periods and under the formative influence of wider social developments.

The goal of this chapter is not to examine Foucauldian theory, or even to attempt to apply it consistently to the analysis of the technologies and practices characterising Web 2.0. Our objective in the following pages is rather to employ the evocative concept of 'technologies of the self' as an object to think with, even if sometimes against or beside the grain of Foucault's work. Thus, we will experiment with the proposition of seeing Web 2.0 media, more specifically blogs and social networking sites, as technologies of the self in the sense implied by Foucault, but without strict allegiance to the nuances of his philosophical analysis.

Foucault defines technologies of the self as a suite of technologies that

> permit individuals to affect by their own means or with the help of others a certain number of operations on their own bodies and souls, thoughts, conduct and way of being, so as to transform themselves in order to attain a certain state of happiness, purity, wisdom, perfection or immortality. (Foucault 1988: 18)

Interestingly enough, if one tries to come up with concrete examples fitting this definition, reading moral tales would be as good a match as body piercing or tattooing. The actual meaning intended by Foucault can be deduced from the historical review of the evolution of technologies of the self that he offers in his treatise. Before heading out to trace that evolution, however, there is another significant point that has to be noted. Technologies of the self should not be conceptualised in isolation, as they are a member of a larger family of technologies including technologies of production, technologies of sign systems and technologies of power. These four types of technologies, Foucault maintains, 'hardly ever function separately' (Foucault 1988: 18), but each plays a key part in the constitution of human beings and requires the individual's specific modification through training and the acquisition of specific skills and attitudes. He admits that perhaps in his previous work he has insisted too much on the technologies of domination and power, but has recently become more and more interested in the 'interaction between oneself and others and in the technologies of individual domination, the history of how an individual acts upon himself, in the technologies of the self' (Foucault 1988: 19). Already in this introduction to the notion, one can sense the tension inherent in it. While technologies of the self invite and presuppose active engagement on the part of the subject and thus come closer to the recognition of agency than any other Foucauldian concept (Sawicki 1996: 174ff), they still retain a strong relationship with domination. They are targeted at 'individual domination' (Foucault 1988: 19) and are closely intertwined with the technologies of domination (of others).

Foucault (1988) starts his historical exploration in 4th–5th century BC Greece with the Socratic tradition, in which young men fulfilled their duty of taking care of the self and thus preparing for public life through ongoing dialogue with their mentors. The Stoic approach which emerged in the 3rd century BC and remained present (if not dominant) until well into the Imperial period saw the loss of dialogue but retention of the mentor/teacher. Rather than engaging in a Socratic dialogue, the individual would spend time in silence, listening to their mentor or listening to public recitations. At that time, a new technology, writing, became a common tool employed in this practice: 'By the Hellenistic age, writing prevailed . . . Taking care of oneself became linked to constant writing activity. The self is something to write about, a theme or object (subject) of writing activity' (Foucault 1988: 27). With the help of writing, the experience of oneself was 'intensified and widened' (Foucault 1988: 28), as attention was paid to nuances of life, mood and reading. Detailed notes of everyday activities and events were taken in order to be examined and contemplated later on for the purpose of self-care. The details of daily life are important here, Foucault points out, because 'they are you—what you thought, what you felt' (Foucault 1988: 29). The mentor would remain the audience and sounding board for this kind of self-elaboration often through the medium of letters. An illuminating example of this practice is the letter by Marcus Aurelius to his mentor Fronto, AD 144–145:

> I slept somewhat late owing to my slight cold, which seems now to have subsided. So from 5 A.M. till 9, I spent the time partly reading some of Cato's Agriculture, partly in writing not quite such wretched stuff, by heavens, as yesterday. Then, after paying my respects to my father, I relieved my throat, I will not say by gargling—though the word gargarisso is, I believe, found in Novius and elsewhere—but by swallowing honey water as far as the gullet and ejecting it again. After easing my throat I went off to my father and attended him at a sacrifice. Then we went to luncheon. What do you think we ate? A wee bit of bread, though I saw others devouring beans, onions, and herrings full of roe. We then worked hard at grape-gathering, and had a good sweat, and were merry and, as the poet says, 'still left some clusters hanging high as gleanings of the vintage.' After six-o'clock we came home. (Quoted in Foucault 1988: 28)

Also significant, in the Hellenistic and Roman period the concern with oneself is not only an educational duty of young people, but also spans one's entire life. The varied set of practices of cultivation of self that emerged in the Greco-Roman period would later be refocused around the objective of 'knowing oneself', which in Christianity becomes a prerequisite for self-denunciation.

Early Christian technologies of the self were centred on one's relationship with God. The self in this case had to be purged and purified through rituals of penitence and self-denial: 'Each person has the duty to know who he is . . . to try to know what is happening inside him, to acknowledge faults, to recognize temptations, to locate desires, and everyone is obliged to disclose these things either to God or to others in the community and hence to bear public or private witness against oneself' (Foucault 1988: 40). The 'symbolic, ritual and theatrical' (Foucault 1988: 43) visual representations of self denial through suffering and (sometimes) death performed by early Christian martyrs were later superseded by the oral confession to a priest or an abbot and the confessional diaries of the monastic tradition. In these confessions, self-examination was directed towards obedience and constant verbalisation and policing of thoughts.

The theme of the technologies of the self is further developed by other participants in the Vermont seminar (1982) of which Foucault's (1988) paper is the centrepiece. The secular counterpart of the monastic tradition that extended into the 16th, 17th and 18th centuries was the 'account book of one's state of sin' kept by Puritans (Paden 1988: 70). It represented another technology of the self intended to ensure permanent monitoring of the soul and the full subordination of thought and deed to church doctrine. From the self-denying Christian who employed technologies of the self to fully subordinate his inner life to God this trajectory can be traced to the man-centred monologues of Hamlet and his generation in which a historically new type of self was being invented (Rothwell 1988); then to the Enlightenment and Rousseau's *Confessions* where one seeks to constitute the self by recounting and reflecting on one's experiences in the secular world (Gutman 1988). In modernity the search was on for the lost 'real self' undertaken in psychoanalysis with its thirst for scientific discovery and belief in basic truths (Hutton 1988).

The idea that practices like these and their pertaining social and technical arrangements could be considered *technologies* of the self contains two important aspects joined in a dialectical opposition. First, technologies are standardised sets of means and rules established under the dominant rationality of a given social order, marked by its inherent power differentials (Feenberg 1999). In this capacity, technologies of the self would have the potential to carry the dominant social and cultural rationality into the heart of self-constitution and thus ensure that the self is shaped in the image and interest of the dominant order. This goes back to their relationship with the technologies of power and domination. At the same time, the notion of technology presupposes conscious and deliberate usage on the part of practitioners, a process that opens up space for agency, subversion and alternative rationality. To the extent that individuals employ the prevailing technologies of the self imaginatively and reflexively, they will be able to take care of the self in a liberated fashion.

Another characteristic of the technologies of the self that stands out in Foucault's discussion is their functioning as a conduit of the interaction between the self and a public of some sort. That public is singular and concrete in the case of the dialogical or epistolary relationship with a master and the confessional disclosure of one's thoughts and feelings in front of the abbot. In some instances, the public is the same self who employs the technology (personal diaries), but at a different point in time and with a different purpose in mind. It is abstract and omniscient in the case of direct sharing with God through dairies of sin. And then it gradually opens up to include the real or imagined audience in the theatre witnessing the Hamletian monologue or the reading public poring over Rousseau's or another author's confessions. It appears that the 'labour of self on self' typically oscillates across a self–other frontier and invites or assumes the gaze of an other.

This observation would not come as a surprise to those familiar with symbolic interactionism which postulates the perennial flow of meaningful symbols between 'I' and others as the main driver of the emergence of 'me', the socialised self (Mead 1934). To think about the constitution of the self as being consciously performed by individuals using socially provided *technologies*, however, takes us beyond the primordial and spontaneous world of symbolic interaction where the 'I' instinctively looks for the reflection of its own image in the eyes of others. Goffman's (1959) dramaturgical model of the 'presentation of the self in everyday life' begins to sketch the contours of conventional practices resembling technology, but it does not go far enough in imbuing these practices with social historicity, material instrumentality and power relations. Nor does it examine carefully enough the nature and origins of the subjective intent that may be at work in them. The expressions given and backstage relaxation of behavioural norms conceptualised by Goffman (1959) appear to be universal and inherent ways in which the self is performed at the archetypical social scene. The concept of technology of the self anchors these and other similar activities simultaneously into the concrete social order and the situated intentions of reflexive agents.

FAST FORWARD TO WEB 2.0

How does the web of rather eccentric concepts and issues spun above connect with the pertinent realities of Web 2.0? Dipping into more contemporary social theory, the current condition of Western civilisation can be described as one necessitating intensive and perpetual labour of self on self. We live in the time of 'do-it-yourself biographies' (Beck 2001) when the self has become a 'reflexive project' (Giddens 1991). Giddens characterises the dynamic interplay between agency and structure in the process of self constitution under the conditions of high modernity thus: The reflexive project of the self 'consists in the sustaining of coherent, yet continuously revised, biographical narratives' and 'takes place in the context of multiple choice as filtered through abstract systems' (Giddens 1991: 5). The more daily life involves a dialectical exchange between the local and the global, the more individuals are 'forced to negotiate lifestyle choices among a diversity of options'. Beck (2001), for his part, depicts a social landscape in which, following the decline of tradition and the social welfare state, the individual has been isolated as the main bearer of responsibility for his or her own fate:

> The ethic of individual self-fulfillment and achievement is the most powerful current in modern society. The choosing, deciding, shaping human being who aspires to be the author of his or her own life, the creator of individual identity, is the central character of our time (Beck 2001: 22)

But then, what are the practices of the self that enable the late modern individual to fulfil this calling? What are the technologies that society places at his or her disposal for achieving that end? Most certainly the technologies of previous eras centred on the family, the church, education and the workplace—on tradition, stable institutions and social welfare—would prove inadequate to the new project of the self. Notably, as Thompson (1995) has argued, the process of self-formation throughout modernity has been also 'increasingly nourished by mediated symbolic materials, greatly expanding the range of options available to individuals' (quoted in Slevin 2000: 157). Typical modern mass media, however, with their centralised and paternalistic organisation fail to equip properly the pursuits of the reflexive self that require 'the processing of contradictory information, dialogue, negotiation, compromise' (Beck 2001: 26) in the process of active management of one's own life.

A late modern medium such as the Internet offers a different base of resources for the active and reflexive shaping of the self. Earlier media brought a stream of events and experiences from around the globe into the local setting of the subject who had to accept or appropriate the presented content. Giddens characterises 'the *intrusion* of distance events into everyday consciousness' (Giddens 1991: 27) as one of the major features of mediated experience in modern times. The interactive nature of the Internet, in contrast, allows a movement that goes the other way round. From within her local setting, the individual can reach into globally distributed resources of information. Importantly, this new kind of mediation makes the local agenda leading in the interaction between local and global. It allows the *elective mobilisation* of distant symbolic resources into everyday consciousness and opens a space for dealing with diversity, contradiction and negotiation.

However, to the extent that Web 1.0 was dominated by solitary, even if self-initiated, practices of searching and drawing on symbolic resources, it offered limited affordance to the practices of the self. It did not support adequately their essential element highlighted in the beginning, namely, the opening oneself up to the scrutiny of a public, of crisscrossing the self–other frontier. The individual remained atomised and isolated even if connected to the proverbial knowledge at his or her fingertips: global symbolic resources available on demand and demand driven by one's own local need and will. There were reactions to atomisation and attempts to reach out to others going on already in the womb of Web 1.0 to be sure. It would be interesting to trace how technologies such as online communities, personal websites and chats led the way to the explosion of do-it-yourself content and conversation media and always-there personal networks which came to be seen as the distinctive features of Web 2.0. For the argument that we are building here, it is important to understand that Internet technology slowly but surely gave rise to a swarm of novel technologies of the self. In the remainder of this chapter we will analyse the potential implications of these technologies and the kind of care for the self they support. 'As there are different forms of care, there are different forms of self', says Foucault (1988: 22). Another question to be explored concerns the interplay between liberty and domination, ethical choice and imposition set in motion by the Web 2.0 technologies of the self. The technologies on which we will focus our examination are blogs and social networking sites.

THE RISE OF THE POPULAR SELF

The fact that Web 2.0 technologies are centred on the self has been noticed and scorned by commentators as Foucault may well have predicted: 'MySpace is about me, me, me, and look at me and look at me' (Fairfax Digital News, cited in Livingstone 2008: 395). As heirs to Christian morality, we tend to see the preoccupation with the self as egoistic and narcissistic, something to discourage rather than foster (Foucault 1988). Equally puzzling and unpalatable is the tendency of the users of Web 2.0 to expose themselves to a wide and undifferentiated audience (boyd 2006; Nussbaum 2007). It is not for the ear or eye of the master, the priest or God himself that these subjects disclose the minute details of their thoughts and deeds. Nor is that disclosure limited to the theatrical floor or an enlightened elite readership. Web 2.0 practitioners strip their souls bare for the ear and eye of anybody who passes by. Of course this is more a theoretical possibility than an empirical reality. At the end of the day, those who pass by are often well-known others, or most typically in the case of blogs, others who share interests, struggles and commitments with the blog's author.[1] Audiences of the performance of the self self-select and respond selectively. Our empirical data along with earlier research (Nardi et al. 2004; Viégas 2005) demonstrate that blog readers form a diffuse network which becomes the authority to which the subject orients its self-making project. The fuzzy boundaries of that network are a source of both risk and opportunity (Livingstone 2008) that the individual at the centre is ready and willing to take in the name of the care of—advancing, expanding or transforming—one's self. As symbolic interactionists have described, the self quivers and changes as it rubs symbolically against other people; however, in this case the process is intentional and reflexive. It is set into motion with the help of a wilfully employed technology. This intentionality does not suggest that subjects always know and can fully control the effects of the operations they perform 'on their own bodies and souls, thoughts, conduct and way of being' (Foucault 1988: 18) using Web 2.0. They are simply keen to engage in experimentation with their selves, something that the late modern condition makes not only possible but also imperative (Beck 2001: 26). At the same time, Web 2.0 practitioners test the powers and limitations of the technologies at their disposal, flex them and bend them inventing new tricks and tactics along the way.

Analysts have already sought and found liberating effects of Internet-based technologies of the self in early studies of virtual communities (Turkle 2005) as well as in the context of Web 2.0 (boyd 2006b). The newly found ability of lay people, previously confined to the status of consumers, to apply their selves to the creation of culture and cultural communities has been noticed and celebrated (Jenkins 2006; Bruns and Jacobs 2006). What has remained relatively understudied is the question of how these technologies might serve as conduits of domination (media descriptions of the 'horrors of the Internet' not withstanding). To address this question, analysis needs to focus on their relationships with the other types of technologies identified by Foucault (1988): technologies of production, technologies of sign systems and technologies of power. Several interesting dynamics can be discerned in this knot of different technologies with potentially high significance for the practices of the self. We will try to propose a few possible routes for investigating these dynamics.

In an era of converging media and 'convergence culture' (Jenkins 2006), Web 2.0 technologies of the self are tightly intertwined with technologies of sign systems such as the mass communication media, the cultural industries and the multi-sensory discourses they propagate. A study of MySpace profiles, for example, documents the function of this site as a stage for the performance of taste (Liu 2007). This is an activity in which the self is being expressed in the terms of the semiotic systems of popular culture: music, film, television, clothing, and so forth. The user interface of MySpace invokes directly the symbols of this culture: your favourite band, film, book, song. The self in MySpace is written not on a clean slate, but through a drop-down menu and a form-filling exercise inviting the individual to map him or herself out along the axes of cultural taste. The widespread practice of bloggers to post links to and excerpts from media texts (Herring et al. 2005) on their personal blogs offers another revealing illustration of the organic, fluent and live connection between Web 2.0 technologies of the self and the prevailing technologies of sign systems such as media and the cultural industry. Technical convergence seems to go hand in hand with ideological convergence, and even though users manipulate the symbols borrowed from mass media and commercial culture relatively freely, their imagination is framed by the properties and semantics of the building material. This certainly is not the only possible scenario for the interaction between these two types of technologies, but its abundance across the blogosphere and social networking platforms is a cause for a pause in the liberation enthusiasm.

The relationship of the Web 2.0 technologies of the self and technologies of production is slightly less transparent and its analysis needs a thoughtful selection of an entry point. We find such an entry point in Terranova's (2004) discussion of 'free labour' in the network society. Networks, as Castells (1996) has argued, constitute the production infrastructure of contemporary Western societies. Once again in this area, technical convergence paves the way towards a closer than ever intertwining of practices of the self with economic production. The labour of self on self, paradoxically, becomes part of the 'free labour', 'pleasurably embraced' and often 'shamelessly exploited', on which the digital economy heavily relies (Terranova 2004: 78). The digital economy, Terranova contends, is a field of experimentation with free/cultural affective labour (Terranova 2004: 79). This experimentation takes many forms spanning the more traditional work-like software development of the open-source movement, on one end, and the largely ephemeral and invisible labour of AOL community-leaders and personal website builders, chats, mailing lists and other amateur forms of production, on the other. Note that the fan fiction and fan community creation so eloquently praised by Jenkins (2006) also fit this bill. Free labour is free in the sense that it is willingly exerted as an outburst of creative energy and self-actualisation. Its shape and substance are not predetermined and imposed by the factory process or managerial plan. It is unpaid labour, and yet is often valorised and channelled into the profit-making enterprise of a capitalist organisation. Terranova's main example is that of the practice of AOL which relied on the commitment and time invested in online community development by numerous volunteer moderators. While the company made tremendous profits in the process, these moderators were never compensated in any way other than gaining prestige among their fellow community participants and a pleasurable feeling of self-fulfilment. With the growth of Web 2.0, a researcher does not need to search for such examples in court proceedings or specialised publications. We live them everyday when we invest time, work and creativity to update our Facebook profile in the morning and hear on the evening news that Facebook founder Mark Zuckerberg is now the third richest person in the world.

Free labour has flooded large sectors of the digital economy. The early experimentation with it has yielded triumphant (for now) business models. Many thousands of bloggers toil away on their computers in the hope that they will attract the number of hits large enough to recommend them as potentially profitable to a company or advertising agency. The self that comes out in these blogs gets appropriately moulded to fit into that project of mass attraction and eventual profitability. On social networking sites users do not need to achieve star status to be profitable to the enterprise. By virtue of employing the technology even if in the least sophisticated way, they lend their profiles and their selves to data mining and network marketing. Thus, re-affirming the Dallas Smythe (1994) dictum, media make money from the inconspicuous work of audiences. Certainly, with the kind of labour they do, immaterial and playful, as it may be, social networking site (SNS) users go a long way further than television audiences who created economic value by simply watching advertising. It may be too harsh to claim that SNS profile makers willingly attune their selves to the commercial machinery. They, however, have very little recourse against the ongoing appropriation of their self-expression into new advertising mechanisms which translate every interest and experience shared by them into a potential consumption trigger. This kind of incorporation is not about capital's descending on authentic culture, Terranova has argued, but represents 'a more immanent process of channeling of collective labour (even as cultural labour) into monetary flows and its structuration within capitalist business practices' (Terranova 2004: 80). Let us be reminded: To the extent that Web 2.0 technologies are technologies of the self (in the hands of the self and for the constitution of the self) the channelling of the labour of their users into monetary flows and capitalist business practices means one thing—the unprecedented blending of technologies of production and technologies of the self. To its credit, digital capitalism has managed to extract profit from the immaterial, but profuse labour of self on self.

Finally we will draw loosely on the 'ludification theory' proposed by Grimes and Feenberg (2009) in the context of online games to discuss the 'rationalising' effect that Web 2.0 technologies have on the practices of the self. Grimes and Feenberg rework a series of theoretical interpretations of play and games to develop a continuum of forms of play starting from the completely free, unstructured and spontaneous playfulness found in everyday life and ending with the strictly rule-governed, highly standardised and often institutionalised games. In many respects, this continuum can be thought of as illustrating the submission of play to technologies of power. Our leading interest in looking for parallels between forms of play and of Web 2.0 use is to establish in what ways and to what extent technologies of power penetrate or emerge amidst the practices of the self furnished by Web 2.0 technologies.

As noted in the discussion of Foucault's texts, the technologies of the self have always contained the element of the other, often an authoritative other for that, whose advice and judgment has presided over the workings of the soul and has offered guidance in the care for the self. In this sense, technologies of the self are always imbued with power, but that power can have different sources and forms. And, importantly, it can be to different degrees disputable and reversible, which for Foucault is the main indicator of liberty (see Foucault 1987). Power could stem from the moral authority of an individual master or spiritual advisor, with whom the subject engages in a nurturing dialogue, or it could be the upshot of a tightly regulated procedure, to which he or she has no choice but to submit. The nature and form of the power present in technologies of the self depend on the way in which these technologies coalesce with society's technologies of power/domination.

In their analysis of online games, Grimes and Feenberg (2009) acknowledge the growing resemblance and interpenetration between production and leisure activities and the numerous ways in which online games give rise to monetary economies and represent sites of free labour. In an interesting next move, they venture to go beyond the registration of this 'economic encroachment' and to study 'how games themselves come to display the same characteristics of rationalization as other institutions of social order and control' (Grimes and Feenberg 2009: 108). 'Rationalisation' here is shorthand for the processes of structuring, stratification, selection and expulsion and the control over subjects that is based on them. Through commodification and technologisation, Grimes and Feenberg argue, games are transformed into yet another realm governed by social rationality. This results in the introduction of strict formal rules, fixed criteria and rigidly defined measurable units designed to frame and control the experience of both players and audience. Such a tightly organised quantifiable experience lends itself readily to commodification:

> As games and play are transformed into an increasingly rationalized set of activities involving huge populations for extended periods, they institutionalize a form of social order. The mass of spectator-players is now organized by the technology of the game much as markets organize consumers, state bureaucracies organize citizens, and production technology organizes workers. (Grimes and Feenberg 2009: 108)

Grimes and Feenberg distinguish several transformations in the historical evolution of play which are also stages through which particular games can pass in their movement from unscripted spontaneity towards rationalisation. The first transition is that from everyday playfulness to organised play. Compare for example students playing on their own in the schoolyard and a game organised for the same students in the same setting, but scripted and led by a supervisor. The second transformation constrains play activities into fixed temporal and spatial conditions. Think of the scheduled practice of beginning hockey players at the local skating rink. The third transition fully transforms a game into a rational system. Play becomes bound by strict rules, quantitative measurements and an institutionalised structure of skills and roles. It is thus ripe for commercial exploitation and profound technical mediation. In the case of online games, the rational structure of the game is programmed into the technical code and player interface and in this way can be tightly enforced.

With this model in mind, we would like to move back to the territory of Web 2.0 and examine the structure of interactions inscribed into social networking sites such as MySpace and Facebook. A logical place to start the application of the model is the larger playing field of social interaction in general. That field hosts a continuum of interactional forms very similar to the one described in the area of play. It includes the everyday chatter of unstructured and spontaneous interactions among fellow men and women; the conventionalised rituals characteristic of particular areas of activity and institutions; the highly structured and tightly regulated relational protocols in formal contexts such as the court of law, the military or the stock exchange; and finally the interaction enwrapped within technological platforms that structure and control exchanges (e.g., various broadcast media formats and groupware designs).

The processes through which the self was constituted vis-à-vis her peer group and the wider circle of people she would call 'friends' used to be located close to the free, spontaneous and authentic pole of this continuum before it was invaded by a horde of new technologies: e-mail, online chats, mobile phones and others.[2] These technologies necessitated some degree of ordering and the evolution of informal protocols (see Ling and Pedersen 2005; Cooper 2001; Ling 1997), but they only affected isolated streams in the overall tangle that made up the interaction among friend couples and groups. They were instrumental technologies with relatively limited application and yet with important effects on the ways relationships were conducted. However, with the widespread use of social networking sites the terrain of friendship was fundamentally transformed. Social networking sites are a hermeneutic technology in the sense introduced by Ihde (1990; see also Bakardjieva 2005) because they insert themselves between the actual world of friendship relations and the person who is an actor in that world. All of a sudden, the landscape of a continent previously experienced firsthand in the course of a multitude of meetings, conversations, gift exchanges, joint activities, and so on, is captured on a technical carrier. The user is invited to perceive and create his or her world of friendships through that technical interface. From this moment on, the technical carrier is going to select and structure the kinds of interactions that constitute friendship. The drop-down menu is going to be the mode through which the self is presented to his or her group of friends; these friends will be counted and quantified in precise ways and the interactions between and among individuals will unfold on a highly visible public stage. Like it or not, the self of the profile holder will be constituted by these different data streams. He or she will be easily compared to others, ranked and placed into categories for different purposes, including marketing. In short, the self will continuously expose its thoughts and deeds to the scrutiny and judgment of its network in the hope to achieve not moral guidance, pardon or salvation, but the major reward that this environment offers, the main currency which animates its moral economy: popularity (our own empirical data; see also Rosenblum 2007; Zywica and Danowski 2008).

This state of affairs in the area of technologically mediated social interaction, then, closely resembles the final stage of transformation of games into systems of social rationality. The same formula that brought games under the strict rule of corporations and bureaucracies is playing out here with still unknown consequences. What is becoming obvious is that friendly social interaction is profitable to site owners and that within this new system of social rationality itself new forms of power are emerging. Among these are the institutionalisation of peer pressure and the stratification of participants on the basis of popularity closely tied to market values and commercial culture stereotypes. To put it figuratively, the guy who went on a tropical vacation and hugged the girls dressed as fashion models shoots up to the top of the ladder. The popular self is the true hero of social networking.

Grimes and Feenberg's (2009) analysis does not stop at the point of extreme rationalisation and commercialisation of play and we do not intend to leave our own argument mired in pessimism either. We know too much about the inventive, devious and subversive tactics of users (de Certeau 1984; Feenberg 1999; Oudshorn and Pinch 2002; Bakardjieva 2005) by now to imagine that system rationality will carry the day unopposed. Even a cursory glance on the empirical usage data that we and others have collected identifies developments that point beyond the prospect of full subjection of the Web 2.0 technologies of the self to the rationality prevalent in their conterminous technologies of production, sign systems and power. As we argued earlier, technologies come to life only in the hands of human operators, practitioners who are situated in a much wider and richer network of relations and experiences than a single technology, no mater how totalising, is able to encompass. To the underlying rationality of the technical script users counter-pose their situated rationalities, intentions and actions (Feenberg 1999). As a result, a variety of user-generated practices anchored in the technology crop up and take hold (Bakardjieva 2005).

As users gain experience and expand their repertoire of practices, the intentional and increasingly skilful employment of Web 2.0 technologies comes to 'informate' (Zuboff 1990) their care of the self. For Zuboff, to 'informate' is the capacity of computers to 'introduce an additional dimension of reflexivity', to produce 'a voice that symbolically renders events, objects and processes so that they become visible, knowable and shareable in a new way' (Zuboff 1988). The hermeneutic quality of social networking sites, for example, allows users to see and interpret their interactions with others from a new perspective. The creative labour that they invest in Web 2.0 technologies and practices helps bloggers and SNS participants to find new affinities and solidarities far outreaching their immediate interaction

Foucault M. 1987. The Ethic of Care for the Self as a Practice of Freedom. In *The Final Foucault*, edited by J. Bernauer and D. Rasmussen. Cambridge, MA: MIT Press.

Foucault, M. 1988. Technologies of the Self. In *Technologies of the Self: A Seminar with Michel Foucault*, edited by L. H. Martin, H. Gutman and P. Hutton. Amherst: The University of Massachusetts Press, 16–49.

Giddens, A. 1991. *Modernity and Self-Identity: Self and Society in the Late Modern Age*. Stanford, CA: Stanford University Press.

Goffman, E. 1959. *The Presentation of Self in Everyday Life*. Garden City, NY: Doubleday.

Grimes, S. and A. Feenberg. 2009. Rationalizing Play: A Critical Theory of Digital Gaming. *The Information Society*, 25(2), 105–118.

Gutman, H. 1988. Rousseau's Confessions: A Technology of the Self. In *Technologies of the Self: A Seminar with Michel Foucault*, edited by L. H. Martin, H. Gutman and P. Hutton. Amherst: The University of Massachusetts Press, 99–120.

Hartmann, M. 2006. The Triple Articulation of ICT: Media as Technological Objects, Symbolic Environments and Individual Texts. In *Domestication of Media and Technology*, edited by T. Berker, M. Hartmann, Y. Punie and K. Ward. Maidenhead and New York: Open University Press, 80–102.

Herring, S., I. Kouper, J. Paolillo, A-L. Scheidt, M. Tyworth, P. Welsch, E. Wright and N. Yu. 2005. Conversations in the Blogosphere: An Analysis 'From the Bottom Up'. In Proceedings of the *38th Hawaii International Conference on System Sciences*, 1–11.

Hutton, P. 1988. Foucault, Freud and the Technologies of the Self. In *Technologies of the Self: A Seminar with Michel Foucault*, edited by L. H. Martin, H. Gutman and P. Hutton. Amherst: The University of Massachusetts Press, 121–144.

Jenkins, H. 2006. *Convergence Culture: Where Old and New Media Collide*. New York: New York University Press.

Jenkins, H. (with R. Puroshotma, K. Clinton, M. Weigel and A. Robison). 2009. *Confronting the Challenges of Participatory Culture: Media Education for the 21st Century*. Cambridge, MA: The MIT Press.

Ling, R. and P.E. Pedersen. 2005. Mobile Communications: Re-Negotiation of the Social Sphere. London: Springer.

Ling R. 1997. One can Talk about Mobile Manners! The Use of Mobile Telephones in Inappropriate Situations. In *Communications on the Move: The Experience of Mobile Telephony in the 1990s*, edited by L. Haddon. COST 248 Report. Farsta: Telia.

Liu, H. 2007. Social Network Profiles as Taste Performances. *Journal of Computer-Mediated Communication*, 13(1). Retrieved from http://jcmc.indiana.edu/vol13/issue1/liu.html. Last accessed 2011–10–04.

Livingstone, S. 2008. Taking Risky Opportunities in Youthful Content Creation: Teenagers' Use of Social Networking Sites for Intimacy, Privacy and Self-expression. *New Media & Society*, 10, 393–411.

Mead, G. H. 1934. Mind, Self, and Society: From the Perspective of a Social Behaviorist. Edited, with an introduction, by C. W. Morris. Chicago: University of Chicago Press.

Nardi, B. A., D.J. Schiano, and M. Gumbrecht. 2004. Blogging as Social Activity, or, Would You Let 900 Million People Read Your Diary? *Proceedings of the 2004 ACM conference on Computer Supported Cooperative Work*, 222–231.

Nussbaum, E. 2007. Say Everything. *New York Magazine*, February 12, 2007. Retrieved from http://nymag.com/news/features/27341/. Last accessed 2011–10–04.

O'Reilly, T. 2005. *What is Web 2.0: Design Patterns and Business Models for the Next Generation of Software*. Retrieved from http://www.oreilly.com/pub/a/oreilly/tim/news/2005/09/30/what-is-web-20.html. Last accessed 2011–10–04.

Oudshoorn, N. and T. Pinch. 2003. *How Users Matter: The Co-construction of Users and Technology*. Cambridge, MA: MIT Press.

Padden, W. 1988. Theaters of Humility and Suspicion: Desert Saints and New England Puritans. In *Technologies of the Self: A Seminar with Michel Foucault*, edited by L. H. Martin, H. Gutman and P. Hutton. Amherst: The University of Massachusetts Press, 64–79.

Rothwell, K. 1988. Hamlet's 'Glass of Fashion': Power, Self and the Reformation. In *Technologies of the Self: A Seminar with Michel Foucault*, edited by L. H. Martin, H. Gutman and P. Hutton. Amherst: The University of Massachusetts Press, 80–98.

Rosenblum, D. 2007. What Anyone Can Know: The Privacy Risks of Social Networking Sites, *IEEE Security and Privacy*, 5(3), 40–49.

Sawicki, J. 1996. Feminism, Foucault, and 'Subjects' of Power and Freedom. In *Feminist Interpretations of Michel Foucault*, edited by S. J. Hekman. University Park: Pennsylvania State University Press, 159–178.

Slevin, J. 2000. *The Internet and Society*. Cambridge: Polity Press.

Smythe, D. W. 1994. Communications: Blindspot of Western Marxism. In *Counterclockwise: Perspectives on Communication*, edited by T. Guback. Boulder, CO: Westview Press, 263–291.

Strozier, R. 2002. *Foucault, Subjectivity and Identity: Historical Constructions of Subject and Self*. Detroit: Wayne State University Press.

Terranova, T. 2004. *Network Culture: Politics for the Information Age*. London and Ann Arbor, MI: Pluto Press.

Thompson, J. 1995. *The Media and Modernity: A Social Theory of the Media*. Cambridge: Polity Press.

Turkle, S. 1995. *Life on the Screen: Identity in the Age of the Internet*. New York: Simon and Schuster.

Viégas, F. B. 2005. Bloggers' Expectations of Privacy and Accountability: An initial Survey. *Journal of Computer-Mediated Communication*, 10(3).

Zuboff, S. 1988. *In the Age of the Smart Machine: The Future of Work and Power*. New York: Basic Books.

Zywica, J., and J. Danowski. 2008. The Faces of Facebookers: Investigating Social Enhancement and Social Compensation Hypotheses; Predicting Facebook™ and Offline Popularity from Sociability and Self-esteem, and Mapping the Meanings of Popularity with Semantic Networks. *Journal of Computer-Mediated Communication*, 14(1), 1–34.

11 Virtual Technologies of the Nation-State

State Administration in Second Life

Stina Bengtsson

Media technologies have played fundamental roles in the birth and maintenance as well as different restructuring processes of nations in diverse ways. New technologies, when they have arrived in the past, have paved the way for the media as tools and arenas for nations to use, and around these technologies media users have also gathered in nationally organised communities. The printing press, landscape painting and the novel have all been used as technologies in the service of nation building and state formation, and the media have accordingly long been one of the central 'cultural technologies of nation-states' (Bolin 2006), where a national media stock is as symbolically important for the formation of a nation as a national anthem, a flag and a national airline (Billig 1995; Kolstø 2000: 243ff). The media also mark the borders of the national territory in terms of coverage, and work as an arena for the national public sphere. Research on the media and nations, historical as well as contemporary, is not surprisingly a vast and rich field of knowledge.[1] In the recent past, this has especially been obvious in relation to the post-Soviet formation of nation-states in Central and Eastern Europe, which are still struggling to 'find the right place on the map' (Jakubowitz 2008: 101ff).[2]

Much of this research has focussed on the traditional mass media: the press, radio and television. In this chapter I will narrow the focus to new, digital and web-based media as technologies of the nation-state, by discussing an advanced example of mediated nation building: the Estonian embassy in the virtual world Second Life. Second Life is a shared 3D online environment, owned and published by the US-based corporation Linden Lab. It was launched in 2003 and has steadily grown ever since, in terms of users and size of digital land. In Second Life users partake as avatars and can buy land where they are free to construct and build environments and buildings, arrange events and happenings, sub-let their buildings and grounds to others, and do basically what they want. The local currency, Linden dollars, is exchangeable to US dollars, and thus to other currencies in the world offline. In 2006–2007 Second Life received immense media hype, which travelled quickly around the world via global media (Kücklich 2009). Mass media all over the world talked about Second Life as 'popular', 'rapidly growing' (Carp 2007) and 'a huge hit' ('Sweden Opens Virtual

Embassy' 2007). Electric Sheep, a dominant 3D producer at the time, predicted in 2007, 'We believe Second Life is following an adoption curve like that of the Web or cell phones, and will hit 100M accounts by the end of 2009'.[3] Today we know that these hopes have not been fulfilled.

But the hype started a worldwide trend. Companies, institutions, churches, organisations, schools and governments entered the virtual world. Five countries established embassies in Second Life. The Estonian embassy was launched in December 2007, the third one in Second Life. It was owned, run and administrated by the Estonian Ministry of Foreign Affairs, and offered services like visa applications and information for foreigners, but mainly worked as a venue for internal cultural events, a national space for Estonians. In June 2011, after nearly four years in business, the Estonian Ministry of Foreign Affairs decided to close the embassy to the benefit of other, more promising, new media platforms.[4]

In the following three sections I will discuss the Estonian embassy in Second Life as a new kind of technology of the nation and the possibilities this new technology can offer the nation, as compared to how the older mass media could work to fulfil this task. First, I will provide an overview of earlier studies of how the media have worked in the service of nation-states. In the second, largest section of this chapter, I will describe and analyse the Estonian virtual embassy and the interplay of technologies and cultures in the creation of this national space. Lastly, I will conclude by discussing the role of new digital and web-based media as a specific technology of the nation-state and in the production of national space.

In the next section three main dimensions of the media as cultural technologies of nation-states will be discussed: first, the role of the media in historical formations of nations; second, its role in the continuous maintenance of nations and national identities; and third, the branding of nations mostly, but not exclusively, used by new and developing nations in the southern and eastern parts of the world. These three dimensions of media technologies in the service of nations all work by means of *representation*—by depicting and constructing images of contemporary society, and by portraying fellow countrymen, addressing the nation as a whole, and representing the nation as seen through the eyes of others.

A main point is to use the Estonian embassy in Second Life as an example of how nations today can *present* themselves symbolically and spatially, be inhabited and resided in by means of new 3D media technologies as virtual worlds—this in contrast to the former mass media representations.

NATION-STATES AND THE MEDIA: REPRESENTING US AND THE OTHER

The printing press, and the cultural transformations that followed it, paved the way for the formation of nations (Eisenstein 1979). Not only because the media long worked as an important arena for nation-building attempts but also, as Benedict Anderson (1983) has argued, because the creation of nations as cultural communities would not have been possible without the media. In his book *Imagined Communities: Reflections of the Origin and Spread of Nationalism*, Anderson puts forward modern capitalism as a fundamental part of early European nation building. He argues that the printing press and the emerging audiences of the novel laid the ground for a sense of community and togetherness among people in a particular territory, sharing a common language.

Some aspects of Anderson's theory are of fundamental importance here. First, the formation of the reading bourgeoisie as a group. The printing press and the spread of novels and newspapers made this emerging social class aware of their likes and of their role in society (Anderson 1983: 83), due to new technology as a means of *representation* of the nation (Anderson 1983: 36). The novel and the newspaper represented the nation to their readers and created a feeling of togetherness among them. This function remains today in, for example, the weather report, which teaches people the names of cities, villages, mountaintops and lighthouses in their nations (as well as in larger geographical regions, e.g., Europe) (Billig 1995; Löfgren 1990). Lastly, Anderson highlights the simultaneous distribution and use of the early mass media. The everyday life structure organised around the distribution of the mass media, and the simultaneous consumption of newspapers, also laid the ground for national imagined communities (Anderson 1983: 44; see also Yoshimi 2003 for a more recent example). This means that the imagined communities of the early mass media grew because people, those who would later come to be citizens of the same nation, suddenly *acknowledged each other.*

Following Anderson's ideas, Michael Billig (1995) has argued that the media today, also in well-established states in Western Europe and the US, uphold nationalism (cf. Yumul and Özkirimli 2000). According to him, the media work by representing the nation in two different ways. First, they 'speak *for* the nation', mainly through the presence of politicians, elected to represent the nation as a whole. Second, the media 'speak *to* the nation', that is, *address* the nation and thus construct it as an entirety (Billig 1995: 98; cf. Dhoest 2007). They can also portray the nation by showing how it looks through the eyes of others (Billig 1995: 100; cf. Tzanelli 2006), as well as by strengthening 'us' by constructing and representing 'them'. This is particularly observable in certain parts of the media like the sports pages and international news, but it is also a more general aspect of nation building in mass media today, as in so-called media events (Dayan and Katz 1992; Billig 1995: 125; Intorn 2002; Stankovic 2004; Reijnders 2007; Hutchings 2008; Mihelj 2008; Perkins 2010). Banal nationalism thus works by representation, in accordance with Anderson's analysis, by depicting and thus defining both the national community in question as well as the people outside the boundaries of the nation-state.

A more recent way to use the media as a technology of the nation is the branding (and rebranding) of nations. In Eastern Europe we have seen many nation branding campaigns during the past decade: Estonia's 'Positively Transforming' in 2002, 'Simply Surprising' in Romania 2003 and 'Easy to Find' in Bulgaria 2005, but the British 'Cool Britannia' campaign of the millennium shift exemplifies the variety of nations involved in this work, whereby even 'old' nations seemingly have a need for rebranding.

Nation branding can be understood as a way to offer a nation as a commodity, to attract tourists and financial investors, and it therefore mainly addresses audiences abroad (Bolin and Ståhlberg 2010).[5] Nation branding works by representing the nation and its people in commercials, slogans and specific colour palettes, at times also connected to particular events (such as the Eurovision Song Contest in Estonia 2002 or the Olympic games in Beijing in 2008). The distinctiveness of these representations (although this term may seem contradictory here), compared with the media representations discussed earlier, is their lack of character and specificity, their absence of history and politics. They aim at attracting investors (and tourists) from other parts of the world by presenting a nation just like anyone else, without obstacles or cultural hinders. Therefore, nation branding campaigns present a nation and its citizens which could belong anywhere, or at least anywhere in the Western world, with the same kinds of blondes, playing golf, shopping in large malls or eating ice cream with their happy children.

To summarise: The building, maintenance and branding of nations work mainly by representation—that of other fellow countrymen in the same nation, a people's own nation viewed through the eyes of others, or by constructing an 'us' and a 'them', like in the sports pages and international news.

As stated, a vast majority of these studies are based on analyses of traditional mass media.[6] However, the past couple of decades have seen rapid shifts in the media landscape, with the Web 2.0 technologies, social networking media and the wider spread of usage of virtual worlds and computer games (cf. Castronova 2005, 2007; Boellstorff 2008). In the next section I will describe and analyse the Estonian embassy as engaged in this new media environment.

A VIRTUAL NATIONAL SPACE

When the Estonian virtual embassy was launched in December 2007, it received a great deal of media coverage internationally as well as locally. The Estonian Ministry of Foreign Affairs initiated the project, but the PR bureau Hill & Knowlton executed it. According to Marten Kokk, Secretary General at the Estonian Ministry of Foreign Affairs, the embassy was part of an intention to use new social networking media to change the image of the Ministry of Foreign Affairs, regarded among many (young) Estonians as grey and bureaucratic.[7] Three basic functions were described in the press release from the inauguration:

> The first is to help Estonia become familiar with complex virtual worlds, which will enhance the nation's ability to use them for disseminating information. The second is to organize high-level meetings, lectures and seminars for diplomatic experts and offer the resulting information to the wider public. And the third function is to introduce Estonia's advancements in the information technology realm to all interested parties.[8]

Successively, its most important function was as a stage for cultural and promotional events, but according to Maria Belovas it was also used as a platform for educational projects, referring to a collaboration with Tallinn University in 2009.[9] The media's attention has also been important from the very start.

The Estonian virtual embassy collaborated with other institutions and private initiators such as museums, universities, Estonian embassies offline and (Estonian) enthusiasts in Second Life when arranging events and happenings. It also encouraged Estonians to visit the virtual world in different ways. For example, visitors were offered two guest avatars at the Foreign Ministry's web page so that Estonians (as well as others) could enter the virtual world without having to create an avatar of their own. Second, land was offered to visitors on the Foreign Ministry's sim for Estonians in Second Life. And third, on May 10, 2008, the European Commission arranged an Open House Day in Tallinn at which they especially informed visitors about the virtual embassy.

The virtual embassy successively developed a more national character, observable in three distinct ways. First, in the change of language: More and more of the information on the embassy building, and on the arranged events, changed from English to Estonian. Second, its temporal structure: Although Second Life generally adjusts to California time, noticeable in a watch visible when in-world, the Estonian embassy followed Estonian time in terms of opening hours and event schedule. This can be compared with the Swedish embassy in Second Life (The Second House of Sweden), which adjusted to visitors from other parts of the world (see Bengtsson 2011). And third, the character of the events, which successively became more internal and nationally oriented, in terms of both partakers (Tallinn University, an Estonian punk band, a famous Estonian sport star, an Estonian diplomat, etc.) and kinds of events (e.g., a discussion about Estonian national identity). The phone number on the events board at the entrance to the building was presented without a national prefix, and the last exhibition shown was about Estonian identity, in Estonian, although it thematised Estonia's relation to the European Union.

The Estonian embassy had a meeting room, an exhibition floor and a desk to welcome visitors. It also had a technology room, especially designed to highlight future technology and communication and to introduce Estonia as an e-nation. It contained three sculptures devoted to 1) Estonian e-voting, 2) the future of communication and 3) cyber war. The communication sculpture told five stories about future human communication, written by the PR bureau Hill & Knowlton. An Estonian dog, the 'news beagle', read

Figure 11.1 Mr. Head telling stories about the future of communication. Photo by author.

Figure 11.2 The 'newsbeagle'. Reproduced with permission of Daniel Vaarik.

Estonian news to visitors, and music from the Estonian radio channel Raadio 2 could be heard in the building.

Second Life is constructed of islands, and the Estonian virtual island was relatively small. As the 'geographic proximity' of virtual worlds (Kalay and Marx 2005) is not visually organised, the view from the Estonian island was limited to the wide-open sky, the blue sea and the horizon.

The architecture of the virtual embassy was ultramodern—a building made of virtual concrete and glass—and in 2008 the embassy was awarded the readers' prize in the 'Concrete Building of the Year' contest, arranged by the Estonian daily newspaper *Eesti Päevaleht*. Apart from a slight resemblance to KuMu, the new Estonian art museum (as well as to art museums in other parts of the world), the architecture was not traditionally 'Estonian', but rather more futuristic and urban. It did not adjust to the laws of nature (although it might have been possible to build offline). This building, in contrast to many other buildings in Second Life (Kalay and Marx 2005; Glazer 2005), did not mimic anything in offline Estonia or other parts of the offline world. According to Daniel Vaarik, former CEO at Hill & Knowlton, and Marten Kokk at the Ministry of Foreign Affairs, the embassy's exterior and interior design was made to reflect the Estonian soul.[10]

Having walked through the thin glass doors of the embassy building, visitors found a spacious and institution-like interior, dominated by the colour grey, large stairs, and most notably the bareness of the walls and, also here, an ignoring of offline circumstances, for example, the lack of handrails along the stairs (since avatars do not get hurt if they fall). The patterns on the furniture in the meeting room and the carpet in the reception room were made with inspiration from old Estonian wooden carvings.

Figure 11.3 Estonian virtual embassy exterior. Photo by author.

Figure 11.4 Estonian virtual embassy interior. Photo by author.

A few examples contradicted the emptiness of the grey walls: a small number of art pieces on the walls, the colours of the Estonian flag tinted on the door handles at the main entrance and a picture of the Estonian president on one of the embassy walls. This national space had three characteristics: It was isolated, playful and fictional, and self-reflecting.

The *isolation* of the Estonian virtual island had to do with the lack of connections and links to contemporary (Estonian as well as other) society, and was expressed both spatially and temporally. As mentioned, the geography of the virtual world in itself produces a kind of isolation that is equal to every space in Second Life. The digital geography limits the view so that no other islands or grounds can be seen or even sighted anywhere. The horizon is not where new visitors appear; instead they materialise out of the sky when they teleport to a virtual island.

More interesting is the lack of paths, which would have made it possible to move from the virtual island to other places. No links were offered to other places in the virtual world, and there were also few links to places in the offline world. The exceptions to this were 1) an information board in the exhibition hall, where visitors could link to web pages referring to places in contemporary Estonia;[11] 2) a message at the front desk saying 'Estonia Republic is offline. Click here to leave a message';[12] and 3) an information board outside the building that showed the name and phone number of the person at the ministry administrating the embassy. In the large building, these links did not dominate the space.

The time references in the embassy also avoided the present, referring either to a mythological national past by presenting a long (and solid) history of the Estonian nation, relating to the mythical singing revolution that led to national independence, through art that thematised Estonian nature

Figure 11.5 Information board with links. Photo by author.

and (folk) culture, and by using traditional patterns in the interior decoration. Or they referred to a high-tech future, as in the sculpture Mr. Head's five stories about the future of communication and in the futuristic architecture of the embassy building. The virtual architecture, and other dimensions of the presentation of Estonia in the embassy, exposed Estonia as a forward-looking nation, with a long and nationally solid past but with little connection to contemporary Estonia.

Figure 11.6 Advertising the motion picture *The Singing Revolution*. Photo by author.

Besides the lack of references to contemporary society already described, it is also interesting to look deeper into *the playful and fictional* character of the national space. Several exhibitions were shown at the embassy during its operating years; the last was an exhibition of Estonian comic strips from daily newspapers discussing Estonia's membership in the European Union and the Euro.[13] But there were also more permanent installations. Above the stairs in the technology room stood a poster promoting a fiction film *The Singing Revolution*, about Estonia's way to national independence. Other playful narratives about Estonia were told by the futuristic sculpture Mr. Head in the technology room. All of Mr. Head's stories thematised means of communication in the future, and some of these stories especially concerned Estonian culture and (imagined) national identity. A good example of this is the story of the Estonian virtual choir, said to be made possible in the future through new technology and available via the Internet (choir singing is one of the most significant features of Estonian national identity, and a key ingredient in the 'singing revolution' that led to independence). Part of this fictional construction of the Estonian soul was also the Estonian beagle, barking and drawing attention to Estonian news. Outside the building hung a miniature of the Earth, alone in the sky.

Michael Billig (1995) argues that traditional mass media work as mirrors for nations, reflecting their self-understanding through the eyes of others. The virtual embassy worked in many ways as a mirror as well, although not one that reflected its observer, or contemporary Estonia (as argued by Kalay and Marx 2005; Glazer 2005). Instead, it worked as a mirror that reflected itself: the embassy building as a national space. The *self-reflective* character of this national space was visible not only in the connections with

Figure 11.7 Estonian virtual embassy exterior. Photo by author.

the future of technology and communication in the sculptures and stories in the technology room, but also in the 'outdoor' pictures of the embassy that covered its interior walls. There were photographs of the embassy in daylight and at night, from several different angles. The virtual embassy was thus not a 'positivistic' reflection of the 'real world', as virtual worlds are sometimes understood (Qvortrup 2002). It did not mirror the world outside or contemporary Estonia. Instead it mirrored itself, and the imagined national space it constituted.

Culture and Technology of Nation-States

There can be many reasons why the Estonian Ministry of Foreign Affairs wanted a virtual national space. They might have wanted to offer the many Estonian expatriated citizens a place to gather virtually. The national independence, and the fact that Estonians abroad are no longer regarded as refugees, has led to the closing of several Estonian houses in various countries around Europe. There can also be reasons to detach it from contemporary Estonian society; through the virtual world, giant leaps can be made into the future.

Whatever the reasons behind the decision to create a virtual embassy in this way, we can conclude that the national space produced here had a very distinct character. It spoke the language of the future in its brave and idiosyncratic architecture, and in the sculptures and futurist stories in the technology room.

To get back to the role(s) of the media in nation building—what kind of national space is discussed here? It is not based on a representation of the people of the nation or an identification based on likeness, as in Anderson's (1983) imagined communities. It is not based on distinction from others, as in the banal nationalism of traditional mass media (cf. Billig 1995). It does not borrow the eyes of other nations to reflect itself from the outside and to boost its own self-confidence. It does not present an indistinct, characterless and commoditised image as in nation branding campaigns (Bolin 2006; Jansen 2008; Bolin and Ståhlberg 2010; Kaneva and Popescu 2011; etc.).

Many dimensions of both culture and technology have to be considered in trying to understand this new kind of nation-based media. On the one hand, culture is there as a background, out of which technology can be used. On the other hand, technology, and the virtual world in this specific case, frames the cultural expressions and their possibilities and limitations.

The Cultural Use of Technology

Up to this point I have argued that the national space constructed by the Estonian virtual embassy was primarily symbolic, an imagined national space, mainly lacking representations of contemporary Estonia. Two cultural frames worked in parallel here: Estonian culture and the culture of the virtual world. These different cultures were active both in the framing of the project and in the use of technology in the construction, presentation and maintenance of the virtual embassy.

The first and most obvious dimension of the culture of Second Life is the playfulness and creativity many researchers have suggested to be the most distinctive characteristics of virtual worlds (cf. Kücklich 2009; Pearce and Artemesia 2009; Castronova 2007). There is thus a greater acceptance regarding the activities, even of state institutions, than there would have been in an offline environment. The barking dog presenting Estonian news is an example of this playfulness, as well as other constructed manifestations of Estonianness in the embassy (like the sculptures, the mini-Earth floating in the air outside the building, etc).

The presentation of the nation in the embassy building referred to Estonian culture and symbols. On the one hand, these signs worked as a subtle promotion of Estonia (as in the relation between the blue sculpture, Mr. Head, and the partly Estonian world-famous corporation Skype); on the other hand, they were experimenting with the possibilities of technology as much as they were truly communicating something to the visitors. Daniel Vaarik searched for fine examples of traditional Estonian patterns to influence the furniture designer, and sent her pictures of some wooden chairs at the University of Tallinn. The carvings were used as models for the carpet in the reception room, although Vaarik has his doubts as to whether this way of using cultural signs was meaningful to those who saw the carpet.

> It fits very well and if you really start looking into these details they really are very cool. The chairs and everything, they really are very good. But of course the average visitor doesn't understand that. They have really low knowledge about what Second Life is really about and how you can create things.[14]

To overcome this lack of knowledge of virtual worlds among the average media user, the Ministry of Foreign Affairs also offered two ready-made guest avatars (available on the ministry's web page), so that people could visit Second Life and the embassy without having to create their own avatar.[15]

Although building, removing, decorating and redecorating are fundamental parts of the culture of virtual worlds, the embassy stayed basically the same during its four years of existence. Some things (such as the barking dog) stopped working and were removed, and different exhibitions were shown there, but otherwise it stayed relatively the same. According to Maria Belovas, resistance to remaking the embassy was considerable within the Ministry; something she asserts has to do with a lack of understanding of the nature of virtuality. Although several people at the ministry identified problems with the embassy project (such as low visitor numbers, too much work involved with creating events, and increasing bad publicity for Second Life), resistance to restructuring or changing the format of the embassy was strong:

> One of my ideas was that we should take down the house, but the thing is that there was an architect who actually built the house and it cost a lot of money. And of course I wasn't there at the beginning, but if you tell people who were there in the beginning and helped with all the projects and you say we should take it all down, you know, they actually think it's a real house. But for me that's the beauty of it, of the virtual world, that you can build something and you can remove it, put it away to the side. You don't have to delete it or anything.[16]

Despite the intention to experiment with new technology when building the virtual embassy, it was in many ways administrated and run like an ordinary state institution offline.

The Technological Frames of Culture

The relation between culture and technology also worked the reverse way: The technology interfered with the construction and presentation of the cultural signs.

Constructing a building in a virtual world is different from doing so in physical reality. No physical laws limit the imagination, although many have pointed out that virtual architecture often mimics the construction and design of buildings offline (Qvortrup 2002; Kalay and Marx 2005; Glazer 2005). This was both an advantage and a disadvantage from the perspective of the Ministry of Foreign Affairs. On the one hand, it offered possibilities to construct a national space where the limits of the offline world did not have to be fully considered:

> So, but taking Estonia, in this position it is natural that we want to be this technology, tech-savvy country. And besides that we're a small country, we don't have a very great climate, we're a small economy and we don't have much to brag about. We're just in a situation where we have some things that we can do, and since this is . . . In the nineties when Estonia was created and we were this . . . I don't know, tech-savvy country and we had this e-government that is also embodied by these rooms with their flashy screens.[17]

The initial lack of experience in this area was naturally a disadvantage, leaving many of the important decisions to the architect. According to both Daniel Vaarik and Marten Kokk, architect Scope Cleaver, who had never been to Estonia but based his impressions of the country upon other media representations, constructed the building based on these impressions and adjusted them to fit the virtual environment. The result is the culturally advanced construction that communicates the front-edge character of the modern, tech-savvy nation.[18]

> The architect . . . he was actually suggesting what it should look like. What, from his point of view, should reflect Estonia as a country. And he made it very modern, so . . . We didn't have to have the technical competence to say what is possible or such. It was very much Scope Cleaver who suggested how it should look.[19]

The embassy building was very large when it came back from the architect, and although the Estonian Ministry of Foreign Affairs wanted a vast and institution-like building, the virtual house turned out to be larger than expected and they realised that these vast spaces could not be left empty. The budget was rather strained, so they tried to come up with ideas of how to fill the empty spaces without overly high costs. The sculptures in the technology room did not cost much, but filled the empty spaces and communicated the preferred Estonianness to the visitors. The self-referential character of the embassy, with photographs of the embassy outside covering its interior walls, may also have been a consequence of this.

The technology of the virtual environment, beneficial to visual and oral communication compared with written information, is the base for the framing of the national culture (as well as the playfulness of virtual culture). The Estonian beagle is thus a kind of hyperlink, albeit adjusted to the playfulness and technological frame of the virtual world.

> The logic of Second Life is different and a visual environment like that does not allow you to write long texts or anything. So we came out with the idea that we'd have a dog . . . In HTML you have hyperlinks which draw attention by being blue, but in 3D worlds the hyperlinks have to draw attention to themselves even more aggressively. So we had to make the hyperlink look like a dog, and bark, in order to focus attention on him. So basically the dog just read a short summary of the news and then there was a link to a homepage or something. So it is actually a hyperlink taking some kind of animated form.[20]

But although the technology seemed unlimited when the embassy was being constructed, the technological restrictions were quite heavy in trying to use it (cf. Gordon and Koo 2008). Due to capacity limits of the technology, only some fifty to sixty people can gather at the same place at the same time in Second Life (cf. Boellstorff 2008). Although the events arranged in the embassy continued to receive quite heavy attention in Estonian media during the entire embassy era, these limits in visitor numbers very soon turned out to be a problem for the Ministry of Foreign Affairs.[21]

> But what I think does work are the events, because we get a lot of crowds together at the same time, we get publicity outside the environment as well. But there are limits, we can only have some fifty-sixty people at a time and well, for me, for example organizing an event

> takes maybe a month of my time, which is to entertain fifty people: I'm not sure that's the way it should be.[22]

These limitations of Second Life, combined with the lack of interest in the virtual world from the larger crowds of users, finally led to the closing of the Estonian virtual embassy.

Producing a National Space in a Virtual World

So, what did the use of new 3D technology mean for the production of the nation in the media? As I have argued, mass media in the service of nations have traditionally worked mainly through representation of the nation's inhabitants and of the national geography (as in the weather report), but also of others, outside the national borders, from whose perspectives the nation's own identity is constructed. There have also been idealised representations, as in the presentation of modern (though characterless) lifestyles in the nation branding campaigns. It is clear that representation, previously so fundamental to the relation between nations and the media, is not so obvious here.

The branding of nations, especially of newly formed nation-states with an urge to attract investors and tourists from abroad and to present themselves to the world, is a phenomenon we have seen a great deal of during the past couple of years. The 2002 Brand Estonia campaign is an early example, and has been followed by many others during the past decade. Several aspects link the Estonian embassy in Second Life to the branding of nations in general and the Estonian branding campaign more particularly. The framing of the Estonian nation as tech-savvy, modern and futuristic is well known from the 2002 Brand Estonia campaign (cf. Bolin 2006; Jansen 2008), and is also similar to campaigns in other countries and parts of the world. The virtual embassy not only presented the Estonian nation in the same kind of favourable light, but also related to famous brands with connections to the nation, although the implicitness of the embassy made it a far subtler act of promotion than branding campaigns usually are.

What made it more unique in this respect, then, is its transformation over the years, whereby the initially internationally oriented embassy turned to new addressees—from foreigners to Estonians—visible in the language used, the character of the events at the embassy and the time structure used on the island.

Also unique was the character of this space. I have argued that the 3D technology of Second Life made it possible for the Estonian Ministry of Foreign Affairs to bypass the earlier representational dimensions of the technologies of nation-states and present the national space differently. Apart from being spatial, a true building to walk into with floors, stairs and ceiling, its distinctive character also followed the temporal bases of the items and decorations in the embassy, where the past and the future of

Estonia(n identity) were more strongly themed by the national symbols in the building than the contemporary nation-state itself. Another dimension of this was the use of playful and mainly fictional symbols of Estonian identity, for example the sculptures, cartoons and fiction film, distinctively different from the constructed naturalism of other branding campaigns. The Estonian embassy in Second Life thus had a stronger character of *building* (rather than branding) a nation, although the playfulness and fictional character of the construction differed from the seriosity of other nation-building symbols (such as a national anthem, a flag, etc.). It was the Estonian national character, the soul of the nation, that was constructed here—not its outside, surface or 'appearance'. The Estonian embassy in Second Life offered a national space for Estonians virtually *in* the media, based upon symbolic construction instead of symbolic representation of the nation; a space for Estonians, but one that could be embarked on, entered and resided in.

NOTES

1. Examples include Eisenstein (1979), Anderson (1983), Billig (1995), Löfgren (1990), Dhoest (2007), Tzanelli (2006), Yoshimi (2003), Intorn (2002), Reijnders (2007), Mihelj (2008), Perkins (2010), Khatib (2006), Roy (2007), Eriksen (2007), Cunningham and Sinclair (2001), Georgiou (2006), Franklin (2004), Hiller and Franz (2004), Parham (2004), etc.
2. Examples on research on Central and Eastern Europe include Bolin (2006), Bolin and Ståhlberg (2010), Jansen (2008), Kaneva and Popescu (2011), Kolstø (2000), Hutchings (2008) and Stancovic (2004).
3. Information material by Electric Sheep, unpublished. Provided by the Swedish Institute.
4. Maria Belovas, information specialist at the Ministry of Foreign Affairs and responsible for administrating the embassy during its last active years, e-mail interview, September 1, 2011.
5. Although others argue that nation branding contains a double articulation striving to change the image of the branded nation abroad as well as internally among its citizens (Kaneva and Popescu 2011).
6. There is also a large field of research concerned with national identities in the Internet age. Thomas Hylland Eriksen (2007), among others, has pointed out that the Internet contradictorily seems to have strengthened national identities, even among Diaspora cultures, refugees and those lacking a geographical national space (Cunningham and Sinclair 2001; Georgiou 2006; Franklin 2004; Hiller and Franz 2004; Parham 2004; etc.).
7. Marten Kokk, personal interview, December 9, 2009.
8. Press release from the inauguration of the Estonian embassy in Second Life, by the Estonian Ministry of Foreign Affairs.
9. E-mail Maria Belovas, Information specialist and responsible for administrating the embassy during its last active years, March 1, 2010.
10. Daniel Vaarik, personal interview, December 9, 2009; Marten Kokk, personal interview, December 9, 2009.
11. The English web page of the the Estonian National Museum, http://www.erm.ee/ (last accessed 2011–09–15), the English web page of Tallinn 2011, http://www.tallinn2011.ee/eng (last accessed 2011–09–15), a page called

Visit Estonia, http://www.visitestonia.com/sv/ (last accessed 2011–09–15) and the web page of the ninetieth birthday of the Estonian republic in 2008, http://www.eesti90.ee/?lang=en (last accessed 2011–09–15).

12. Estonia Republic was one of the Foreign Ministry's two official avatars.
13. Estonia entered the European Union in 2004 and changed its currency to the Euro in January 2011.
14. Daniel Vaarik, personal interview, December 9, 2009.
15. Marten Kokk, personal interview December 9, 2009.
16. Maria Belovas, personal interview, April 14, 2010.
17. Daniel Vaarik, personal interview December 9, 2009.
18. Marten Kokk, personal interview, December 9, 2009; Daniel Vaarik, personal interview, December 9, 2009.
19. Marten Kokk, personal interview, December 9, 2009.
20. Daniel Vaarik, personal interview, December 9, 2009.
21. Also mentioned by Marten Kokk, personal interview, December 9, 2009.
22. Maria Belovas, personal interview, April 14, 2010.

REFERENCES

Anderson, B. 1983. *Imagined Communities: Reflections on the Origins and Spread of Nationalism* (2nd ed.). London: Verso.

Bengtsson, S. 2011. Virtual Nation Branding: The Swedish Embassy in Second Life. *Journal of Virtual Worlds Research*, 4(2), 1–26.

Boellstorff, T. 2008. *Coming of Age in Second Life*. Princeton, NJ: Princeton University Press.

Billig, M. 1995. *Banal Nationalism*. London: Sage.

Bolin, G. 2006. Visions of Europe. Cultural Technologies of Nation States. *International Journal of Cultural Studies*, 9(2), 189–206.

Bolin, G. and P. Ståhlberg. 2010. Between Community and Commodity: Nationalism and Nation Branding. In *Communicating the Nation: National Topographies of Global Media Landscapes*, edited by A. Roosvall and I. Solivara Moring. Göteborg: Nordicom, 79–101.

Carp, O. 2007. Svensk ambassad i Second Life. *Dagens Nyheter*, January 25. Retrieved from http://www.dn.se/kultur-noje/svensk-ambassad-i-second-life-1.442651. Last accessed 2011–09–15.

Castronova, E. 2005. *Synthetic Worlds: The Business and Culture of Online Worlds*. Chicago: University of Chicago Press.

Castronova, E. 2007. *Exodus to the Virtual World: How Online Fun Is Changing the World*. Basingstoke, UK: Palgrave Macmillan.

Cunningham, S. and J. Sinclair. Eds. 2001. *Floating Lives: The Media and Asian Diasporas*. Lanham, MD: Rowman & Littlefields Publisher Inc.

Dayan, D. and E. Katz. 1992. *Media Events: The Live Broadcasting of History*. Cambridge, MA: Harvard University Press.

Dhoest A. 2007. Identifying With the Nation: Viewer Memories of Flemish TV Fiction. *European Journal of Cultural Studies*, 10(1), 55–73.

Eisenstein, E. 1979. *The Printing Press as an Agent of Change: Communications and Cultural Transformations in Early-Modern Europe*. Cambridge: Cambridge University Press.

Eriksen, T.H. 2007. Nationalism and the Internet. *Nations and Nationalism*, 13(1), 1–17.

Georgiou, M. 2006. Diaspora, Identity and the Media: Diasporic Transnationalism and Mediated Spatialities. Cresskill, NJ: Hampton Press.

Glazer, N. 2005. Architecture and the Virtual World. *First Monday*, 11(2). Retrieved from http://firstmonday.org/issues/issue11_2b/glazer/index.html. Last accessed 2011-09-25.

Gordon, E. and G. Koo 2008. Placeworlds: Using Virtual Worlds to Foster Civic Engagement. *Space and Culture*, 11(3), 204–221.

Franklin, M. 2004. *Postcolonial Politics, the Internet, and Everyday Life: Pacific Traversals Online*. London: Routledge.

Hannerz, U. ed. 1990. *Medier och kulturer*. Stockholm: Carlssons.

Hiller, H. H. and T. M. Franz. 2004. New Ties, Old Ties and Lost Ties: The Use of the Internet in Diaspora. New Media & Society, 6(6), 731–752.

Hutchings S.C. 2008. Saint Petersburg 300: Television and the Invention of a Russian (Media) Tradition. *Television & New Media*, 9(1), 3–23

Inthorn S. (2002). The Death of the Hun? National Identity and German Press Coverage of the 1998 Football World Cup. *European Journal of Cultural Studies*, 5(1), 49–68.

Jakubowitz, K. 2008. Finding the Right Place on the Map: Prospects of Public Service Broadcasting in Post-Communist Countries. In *Finding the Right Place on the Map: Central and Eastern European Media Change in a Global Perspective*, edited by K. Jakubowitz and M. Sükösd. Bristol: Intellect, 101–124.

Jansen, S. C. 2008. Designer Nations: Neo-liberal Nation Branding—Brand Estonia. *Social Identities*, 14(1), 121–142.

Kalay, Y. E. and J. Marx 2005. Architecture and the Internet: Designing Places in Cyberspace. *First Monday*, Special Issue 5: Virtual Architecture at State of Play III. Retrieved from http://www.firstmonday.org/htbin/cgiwrap/bin/ojs/index.php/fm/article/viewArticle/1563/1478. Last accessed 2011-09-25.

Kaneva, N. and D. Popescu 2011. National Identity Lite: Nation Branding in Post-Communist Romania and Bulgaria. *International Journal of Cultural Studies*, 14(2), 191–207.

Khatib L. 2006. Nationalism and Otherness: The Representation of Islamic Fundamentalism in Egyptian Cinema. *European Journal of Cultural Studies*, 9(1), 63–80.

Kolstø, P. 2000. *Political Construction Sites: Nation-Building in Russia and the Post-Soviet States*. Oxford: Westview.

Kücklich, J. R. 2009. Virtual Worlds and Their Discontents: Precarious Sovereignty, Governmentality and the Ideology of Play. *Games and Culture*, 4(4), 340–352.

Löfgren, O. 1990. Medierna i nationsbygget; Hur press, radio och TV gjorde Sverige svenskt. In *Medier och kulturer*, edited by U. Hannerz. Stockholm: Carlssons, 85–120.

Mihelj S. 2008. National Media Events: From Displays of Unity to Enactments of Division. *European Journal of Cultural Studies*, 11(4), 471–488.

Parham, A. A. 2004. Diaspora, Community and Communication: Internet Use in Transnational Haiti. *Global Networks*, 4(2), 199–217.

Pearce, C. and Artemesia. 2009. *Communities of Play: Emergent Cultures in Multiplayer Games and Virtual Worlds*. Cambridge, MA: MIT Press.

Perkins C. 2010. The Banality of Boundaries: Performance of the Nation in a Japanese Television Comedy. *Television & New Media*, 11(5), 386–403.

Qvortrup, L. 2002. Cyberspace as Representation of Space Experience: In Defence of a Phenomenological Approach. In *Virtual Space: Spatiality in Virtual Inhabited 3D Worlds*, edited by L. Qvortrup, London: Springer, 5–24.

Reijnders, S. 2007. Media Rituals and Festive Culture: Imagining the Nation in Dutch Television Entertainment. *International Journal of Cultural Studies*, 10(2), 225–242.

Roy, I. S. 2007. Worlds Apart: Nation-Branding on the National Geographic Channel. *Media, Culture & Society,* 29(4), 569–592.
Stankovic, P. 2004. Sport, Nationalism and the Shifting Meanings of Soccer in Slovenia. *European Journal of Cultural Studies*, 7(2), 237–253.
Sweden Opens Virtual Embassy in Second Life. 2007. *Spiegel Online International*, January 30. Retrieved from http://www.spiegel.de/international/0,1518,463073,00.html. Last accessed 2011–09–15.
Tzanelli R. 2006. 'Impossible Is a Fact': Greek Nationalism and International Recognition in Euro 2004. *Media, Culture & Society*, 28(4), 483–503.
Yoshimi S. 2003. Television and Nationalism: Historical Change in the National Domestic TV Formation of Postwar Japan. *European Journal of Cultural Studies*, 6(4), 459–487.
Yumul, A. and U. Özkirimli. 2000. Reproducing the Nation: 'Banal Nationalism' in the Turkish Press. *Media, Culture & Society*, 22(6), 787–804.

12 The Scary Promise of Technology

Developing New Forms of Audience Research

Joke Hermes

It is said that we are heading towards a 'convergence culture' (Jenkins 2008). The post-broadcast era heralds a promise of further democratisation through technology. Of course, there is a problem that technology has thus far hardly delivered its promise of liberation and freedom. Recurring debates set in the perennial modernist dilemma of promise and danger (Jensen 1990) present more than an ideological stranglehold on innovation and emancipation. They reflect fairly accurately that neither technology nor 'media' on their own can or will generate social change sui generis. The crux of the matter is in how media technology is put to use and how we choose to interpret technology. From my point of view technology is a cultural construct, which becomes meaningful in concrete material practices that are themselves shaped in and by relations of power and domination.

In this chapter I will take the perspective of an audience researcher, interested in furthering media literacy and critical reflection. As elsewhere, technology promises to be of immense value to research practice. It is a mere matter of time until Clay Shirky (2008), Steven Johnson (2010) or other internet-and-technology optimists publish a highly readable book that will have the more worrisome intellectual wondering how she can be such a cantankerous pessimist and technophobe as not to immediately use new and better strategies just because they involve machines. After all: Pen and paper are as much a cultural technology as the new data analysis software or the idea that the 'wisdom of crowds' can and should be tapped (Surowiecki 2004). Moving from 'technology and culture' to the interpretation of technology and the technology of interpretation, it will become clear how technology and culture are imprinted on one another but also remain polar opposites, able to spark off old contradictions between cultural pessimism and optimism. Which is not to argue against the use of technology in cultural research but rather to address what is required of dedicated research technology given the culture in which it needs to function. I intend to address this question broadly, going from how technology is perceived, to how it has been studied from a cultural studies point of view (which I share), to why a certain degree of distrust may not be misplaced, even when engaging in a heart-warming utopian vista of the perfect aid to help the audience researcher.

TECHNE, TECHNIQUE, TECHNOLOGY

'Technology' is a scary word. It suggests rationalisation, measurement and results, efficiency. It excludes humaneness, art and emotion. 'Machine' is a much friendlier word, calling forth cars and scooters, wood crafting or kitchen tools, or iPods and DVD recorders. Machines help us do things. They appear not to force us into regimes, routes or routines (while they may well do so). Machines allow for the pleasure of irrational choices. Driving a car today in full knowledge of its environmental effects is a form of madness, or perhaps simply a means to savour the pleasure of being knowingly cognitively dissonant. Machines provide us with luxury, with other and different qualities of experience, with pleasure and an instant shield to allow us to not know the cost at which we use them. This must be the reason there is little as irritating as machines that do not function well. Think of a stapler that crunches its staples or, indeed, techniques of analysis that are difficult to understand and use or fail to deliver results beyond a summary that reflects search strings you first had to devise yourself anyway. The moment a stapler or software becomes 'technology', it is all the more obvious that operating such 'machinery' has to follow rules and procedures. Obligations and responsibilities are suddenly on the map again (staple pages in the right order), and gone is the pleasant and probably utopian wish that things take care of themselves.

The 1973 edition of the *Oxford Concise Dictionary* concurs with Wikipedia that 'technology' is the

> usage and knowledge of tools and crafts, and how it affects a species' ability to control and adapt to its environment. Technology is a term with origins in the Greek *téchnē* (*τέχνη*), 'craft' and *-logía* (*-λογία*), the study of something. 'Technology' can refer to material objects of use to humanity, such as machines, hardware or utensils, but can also encompass broader themes, including systems, methods of organization, and techniques. The term can either be applied generally or to specific areas. (Wikipedia, consulted October 1, 2009)

Media technology is a case in point, and so are analytic techniques.

Technology includes machines, what they do for us (and to us), how we understand the regimes they enable and install but also how we understand and value what machines, techniques and regimes do for us and to us. Given this complex and wide-ranging field that is 'technology', it is hardly strange that feelings about it are characterised by ambivalence. For it seems to me that I am not alone in feeling ambivalent about technology. Again, there is abundant choice in media research when it comes to 'media and communication machines' and their 'cultural' effects—whether it concerns their ability to help human societies control and enable knowledge of and grip on the social (and material) world or how they control us (McLuhan

1964; Postman 1985; Mander 1978; Keen 2007). It is worryingly easy to exaggerate the role of technology, even if this means going much further than these authors do. In everyday talk it makes perfect sense to suggest that technology precedes culture—or even to reduce culture, defined as practices of making sense of the world, to a mere lens or conduit formed within the boundaries and by the logic dictated by technology. In my understanding, culture is as important as technology, both types of practices will enable and restrict each other.

The merest whiff of technological determinism does not go well with understanding culture as a set of practices, including the multilayered results of those practices. It appears to exclude agency. It makes it difficult to understand the variety of practices that occur in relation to media technology. It forces the critic to take a highly abstract point of view. As a result, practices of media production and consumption lose the detail that makes them interesting and meaningful. For those who are fascinated by people, by art and creativity, including everyday arts of making do and the pleasure of going beyond what is strictly necessary in performing daily work, technology as such, on its own, holds little attraction. Although 'logic', 'apparatus' and other terms that refer to understanding how things work (and thus to technology) are a part of most cultural analysts' vocabulary, there is always some distance to the really, seriously technical. Several generations of qualitative media analysts share this sentiment. It has had detrimental effects on our ability to take technology and its promise beyond media technology seriously, whether it regards recognising and weighing the possibility of democratisation or research methodology. The moment Luddites such as myself understand that technology can and should be enjoyed, experienced as the luxury of a well-tuned car or a sharply honed Japanese knife, there will be a better chance of understanding and valuing the complex interrelatedness of culture and technology.

Let us therefore assume that it is not technology but its hard-ass reputation that is the problem. Qualitative media research by necessity slowly made its way into the academy as cultural criticism (Ang 1990; Drotner 1993). Certainly, initially what became known as 'media ethnography' was the realm of new groups entering the academy, who had very little research money at their disposal (Drotner 1993: 28). The listening ear, a pen and a piece of paper were all the technology that very small budgets allowed for. However, budget constraints hardly constituted a strong legitimation to use a method, and a case was built for research methods that could by and large do without (expensive) technology and machines. Qualitative research tried not to impose its own frameworks on informants, instead showing them respect, using their categories and rewarding them by offering quality conversation (McRobbie 1982). This was long before statistical methods became available for home use, where they do not need mega mainframe machines to run. Ironically and unfortunately, qualitative media research found its identity, born at least in part from circumstance, by underlining its difference from quantitative science just a short time before computing power became cheap. Even more ironical is the effect this 'democratisation' of a research tool has had on how quantitative methodology and especially SPSS (Statistical Package for the Social Sciences) are now thought about. While computing technology today is easy to use and fairly cheap, the building and validating of a questionnaire and approaching respondents remain a costly affair. There is more to statistical analysis than pushing the button 'compute'. A real rapprochement between quantitative and qualitative methods is therefore harder and harder to achieve, as their die-hard proponents have their hands full defending the 'real' thing and obtaining research funds to use them.

In the period during which media ethnography was established as a legitimate enterprise in the humanities and social sciences, it would literally not have paid to be interested in the possibilities technology had to offer. Research budgets for new fads such as 'media ethnography' were virtually non-existent, or were in any case smaller than they are now. Ideological barriers grew very high over a short period. Apart from recording devices, technology was problematic, even suspect. The strength of qualitative media research is in its understanding that listening to people is an art; it needs to be taught and learnt and mastered. Its strength is in understanding that it matters who is listening to whom, that we all have our specific, historical locatedness and so on and so on (Hammersley and Atkinson 1983; Gray 2003). We recognise, correctly I still think, that 'objective' methods result in losing sight of the complexity of everyday meaning making. We recognise that methods may veil power relations. We could do more to develop better ones, to go with the steep learning curve societies seem to be on when it comes to handling programmed machines. We could admit that something can be said for robust and systematic data analysis: It could help rather than hinder strong theorisation of empirical work.

This suggests a threefold set of priorities for qualitative audience research. First, technology needs to be rescued from its ambivalent status (from the likes of me). There is important work in the humanities and cultural studies that for some reason I have never really regarded as being 'about technology' while it actually is (see following discussion and, e.g., Caldwell 1995; Kittler 1997). Second, a little more craftsmanship and method could well be used in qualitative data analysis in audience research. With or without the help of machines and software, it is high time that qualitative research found ways to show the systematicity in its analytical stages, moved away from the individual interpretative brilliance or intuition of practitioners and offered its method as a technology, a craft consisting of specific techniques, to reconstruct meaning and to understand practices of meaning making. How do we analyse culture, what do we pay attention to and what do we not, how do we know what specific examples mean? Third, although this is certainly also a question of epistemology, I will look for 'technique' and tools, and especially for what fun there might be in using them. Furthermore, I will try to keep working on different levels: I am not interested in 'mere' methodology but rather in the ensemble of questions around media technology.

INTERPRETING TECHNOLOGY

For inspiration I offer a couple examples of media technology and culture that could, perhaps, provide good starting points for research from the perspective of the audience/user. There is, first, historical work that directly engages with media technology (Kittler 1997; Van Dijck 2007), tending to show how technology shapes our world and enables ways of making sense of it. Caldwell's (1995) study of televisuality fits in this particular set. Caldwell argues the complex determination of style, genre and quality in American television. Digital video as technology enabled a shift that would result in new genres (among them reality television). Cheap non-fiction television tended to be boring; shooting on location is costly, and studio productions presented mainly talking heads. Digital video was revolutionary in that it allowed for all sorts of manipulation of the material, for inserts and split-screen broadcasting. With an explanatory caption, blurry images from fixed cameras in police cars and from handheld cameras became perfectly usable and underlined the authenticity of the footage. A man confessing to adultery becomes that much more exciting to watch when the insert shows his wife's face at another location, of which he is unaware. Reality TV and the talk show quickly became booming business. Digital technology did not do this on its own; it needed the crisis in network television and the discussion of what constitutes 'quality' in television. With a more refined set of terms, what used to be 'poor quality'—unfit for broadcasting—could become material to be worked on, with qualities in its own right, for instance as it was real and almost 'live'.

As with television technology itself: Without the right set of circumstances, technology can fail and be forgotten or used for highly different purposes (Zielinksi 1999; Wieten 1994). While technology can seem to be an autonomous force, it only becomes so when embedded in social and economic power relations. In this regard the fantasies of (prospective) users would be of great interest, as well as their experiences, understood against the background of the actual historical development of particular technologies. An example today would surely be the redefinition of television for mobile, miniaturised personal screens. Conceivably, a decade from now the family living room screen will still be what we call television, while news or quality drama series will be defined by content rather than by their carrier technology. Ultimately, this will vastly change the meaning of 'television'.

Second, Closer to home, for my current purposes, there is the groundbreaking audience research that studied technology via its usage rather than for its own merit. There is David Morley's (1986) *Family Television* in which he showed how the remote control, a piece of technology, extended gendered power relations in households. Or Ann Gray's (1992) work on the introduction of the VCR and how this new machine was likewise slotted in an already existing regime of gendering technology and strengthened it. While women would feel perfectly comfortable handling complicated cookers and washers, the VCR, they felt, was completely beyond them. It is not just a question of performing one's gender adequately (Butler 1990). Gray notes that not being able to programme the VCR was also 'calculated ignorance'. As one of her informants explained, the moment it was known she could handle the machine, programming the thing would be added to her list of chores. Technology in these cases was certainly not a source of democratisation; if anything it confirmed existing regimes.

Technology, clearly, also requires getting used to and practice. It requires a certain measure of self-assurance and should not be too far from the kinds of scripts we tend to follow in everyday life that prescribe what is and is not appropriate behaviour. Although technology often has a 'black box' type of exterior and presents itself as fully fashioned (thus strengthening a determinist interpretation of its social and cultural role), it does adapt. The VCR, and later the DVD, merged with men's and women's lives over time. Arguably, the technology itself has also become more user friendly, which would also explain why few people today would claim that programming a VCR was too difficult a task.

Perhaps more importantly, understanding the place and role of technology in everyday culture requires a reciprocal understanding of the contexts in which it is introduced and used. How are these context shaped, and what rules and notions govern them? How do we know which contexts are relevant? Morley and Gray single out gendered relations in households, but the idea of the single household as a meaningful unit could have been a starting point too. The VCR can then be seen as part of a lineage of machines that produced the single-family household and effectively disciplined the working classes and kept them off the street, as John Hartley (1999) has argued. The fridge and the TV set are his key examples.

Third, of great interest, too, is yet another type of work that benefits from the insights of actor-network theory. Inspired by the work of Bruno Latour, the central tenet is to let go of the distinction between human and other agents and to focus on how an ensemble of actors produces particular end results. From recent work and work in progress I have seen, I make note of Liam Berriman's (2009) research on the BBC cross-media children's programme/game *Bamzooki*. Cross-media genres are in themselves of great interest in this discussion, as the question surfaces as to whether or not the actual medium matters greatly and, if so, how. One of the many things that are interesting in Berriman's work is the melancholy that infuses the way the *Bamzooki* producers talk about television as a medium that is almost obsolete. It certainly does require Internet 'extras' such as a module to build your own 'bamzooki' and to have the possibility for gaming. By pronouncing television yesteryear's medium, the producers make room for funding and energy to be spent on building a website and software, means by which they ultimately hope to control their wayward child audience. Such melancholic talk about television as a screen for singular programmes easily becomes a self-fulfilling prophecy. While the demise of television could be understood as a threat by the producers of *Bamzooki*, they are happy to keep the melancholy mellow and underline the happy talk about new possibilities. This discursive strategy or technique will eventually ensure that they will hold onto their jobs, now redefined as cross-media producers. Incidentally, the melancholic suggestion of the end of television will change the remit of public broadcasting to include the Internet, by its nature more of a 'narrowcast' medium.

Even more clear when it comes to the effectivity of techniques and technologies are the findings of Cindy Peters (2011) at the University of Amsterdam. She looks at a children's website set up and funded by KRO, one of the Dutch public broadcasting organisations. Children are invited to upload self-made movies. Most of the uploaded material is of poor quality and tends to simply imitate existing genres (a commercial, a quiz, a newscast). But *TV Tower* becomes that much more interesting when the material is organised in relation to the role of the camera. In some cases the *camera is merely a toy*: The children's home movies show them making funny faces and so on. The *camera can also replace a missing audience*. While my sister and I would round up our parents and our friends' parents when we wanted to perform circus acts, the camera allows children to perform to absent parents and replace them. In the home movies that clearly imitate familiar genres, the camera becomes almost incidental. The genre or format is given centre stage, and clearly the pleasure of *showing off your television literacy* as a child is important. The few really good clips exhibit both media literacy and creative intervention. These clips can be recognised because they are not filmed as a single shot but consist of more shots, and camera angles are chosen with care. *The camera in such cases becomes a tool*, technology to be enjoyed for what it can help you do.

Cindy Peters' work shows that it is highly relevant from the perspective of audience research to understand the effectivity or productivity of media technology. It gives depth to media literacy and investments in media making on the part of the children who post material on *TV Tower*, and enables the expression of how media technology, screen genres (the home video is also one they like to imitate) and childhood are meaningful categories for children: It is a period in which you practise skills you do not need to have perfected yet. As a child you would like to have an audience but cannot always count on one being available, and so on. Much of this would be hard to express in an interview situation or even in participant observation. In the latter case, the researcher would obviously take the place of the camera as 'ersatz' audience. Technology thus intertwines with and changes social practices (parenting and childcare), informal learning (media literacy) and artistic practice: There is no one single context that makes clear the cultural significance of technology; there is a whole range of them at play at the very same time. Moreover, these are 'articulated' contexts, affecting one another.

It is almost as if technology more easily finds its place if you happen upon it incidentally: your parents' video camera, a friend making dough in an unfamiliar way. When you want technology to work for you, and make true on its promise, it can also be a major stumbling block. In recent research I experimented with participative design in a team project aimed at resulting in a web-based TV series (Hermes 2009). The participants were writers who published short stories on a community website for Dutch-Moroccan youngsters (Marokko.nl). Instead of providing more insight into interactive storytelling and Dutch-Moroccan identity and community building, I think we learned more about how expectations of both television and the web as technologies sabotaged our attempts to keep the project small and manageable. The sheer reputation of television as broadcast technology soon shifted the ambition of the writing team to produce something for national TV, which meant negotiations with a television producer, a funding agency, and so on. Whereas we had counted on the cheap technology and generous acceptance threshold of the community for something coming from their own group, we were suddenly faced with a totally different set of challenges.

From older research as well as these two new examples of research that includes or focuses on the meanings of media technology, it is clear that technology is not in itself destiny (as in technological determinism). It is also clear that naïveté about the possibilities (or dangers) of technology will not do. The Ostrich Doctrine in media research is not an option; it has to be accepted that technology is part of the complex social and cultural relations and constellations that qualitative media and cultural studies research has always engaged in to understand everyday meaning production and the power relations that structure it. Technology is like an ingredient in a recipe: often hardly recognisable in the end product (processes of meaning production) but crucial nonetheless. The challenge now is to rescript media and cultural studies methodology to go beyond accepting technology as part of the field, and also recognise it as a possibly strong ally in data collection and analysis.

Such a rescripting may work in a direction opposite to the engaged critical position media and cultural studies has historically taken up. If done well, it will lend more legitimacy to 'soft' methods, especially in the social sciences. Will it, however, also accord with the interpretative, hermeneutic 'soul' of cultural studies? Being incorporated into 'normal science' can be both a good thing (more funding possibilities) and a spanner in the political project. Interpretation, after all, is never a value or preconception-free exercise. On the contrary, its strength is that it combines understanding of particular situations, states of being, practices or developments in the light of prior knowledge and theory. While we may—and do—need to be more convincing in the presentation and perhaps even in the analysis of research data, repressive tolerance is not what cultural studies needs. Of course, convergence culture, as Jenkins calls it, or Web 2.0 or 3.0, suggests that technology may well become more malleable, easier to manipulate and use. While new media and Internet utopians quite rightly point out society's steep media-related learning curve (Johnson 2005) or the untapped potential that may well be freed under new technological possibilities (Shirky 2008; Johnson 2010), there is the issue that forces of innovation will have to do battle with the economic powers that be. If indeed the availability and popularity of new media technology present a real challenge to dominant social and economic culture, it will find ways to fortify its position. As Henry Jenkins remarked a couple of years ago,

> At the moment we are on a collision course between a new economic and legal culture which encourages monopoly power over cultural mythologies and new technologies, which empower consumers to archive, annotate, appropriate, and re-circulate media images. The recent legal disputes around Napster represent only a skirmish in what is likely to be a decade long war over intellectual property, a war which will determine . . . the nature of creative expression in the 21st century. (Jenkins 2003)

From an everyday-life audience perspective, my somewhat pessimistic sense is that it may well take a long time for significant shifts to occur as a result of new and cheap media technology. In addition, pessimists will point out, based on earlier adaptation of new media forms, it is near impossible to predict what these shifts might be. Corporate business is doing its utmost to control developments, and has been successful to a degree. The technical availability of Web 2.0 and cheap media technology does not mean that either technological skills or a sense of ownership and self-assurance in the face of these new possibilities are an everyday reality, let alone real ownership of new possibilities and technologies. A *naïve* belief in Web 2.0 technology or even a naïve belief in new methods of audience research such as participative design, may well be our worst enemy. Enthusiasm and hope, so easily generated by the hubris that technology also seems to give rise to, need to be tempered by realising our own humble status in the face of the unintended consequences of most any human action; this applies especially to those in research, policy and government (Merton 1949/1968; Bovens and 't Hart 1996). Really, what is needed is a broader set of skills to offset the partly healthy paranoia we already have: sophistication, in other words.

THE TECHNOLOGY OF INTERPRETATION

Neither technological nor interpretative sophistication are easily achieved. From the perspective of audience studies, the future offers a real challenge. Control is illusory—as would be fully comprehending either the nature or the development of the media landscape-to-be. Even if we agree that understanding media technology, the new (forms of the) media text and (new) experiences of media use would do well with more refined techniques and that technology has historically had negative connotations in qualitative media research, there are two other grave dangers that need to be addressed. I have argued earlier (Hermes 2009) that critical audience studies can be a real trap. If the aim of audience studies is to give a voice to groups of audience members, or even to emancipate these groups, why is it mostly the voice of the researcher we hear? How is it that she directs and produces the tale that results from her research? And second, not only is there the danger of becoming a puppeteer; there is also that of over-identifying with audience group preferences rather than staying in critical dialogue mode.

Declining to identify would close the researcher off from understanding. A hermeneutic approach demands a certain measure of empathy. Nor would we want to devise a technology of interpretation that goes back to objectivist standards or relies on 'neutral' techniques. The challenge is to develop engaged techniques, however counterintuitive this may sound if we want to understand the interplay of culture and technology in everyday life and allow for the democratic possibilities of Web 2.0 technology to extend to research practice.

The easy part of this project is to use available techniques and technology to further open research to full dialogue (as in participative design, for instance, or what I have called 'civic research', in which all research participants have a voice in the research design and know how they will benefit from the project; Hermes 2006). Voice recorders and cameras are useful tools, as are dedicated web pages that allow different users to add material or comments without removing the work of others, or those that offer collective work spaces, and so on. Before long, I should think, we will have computer software that will allow for the automatic transcription of spoken text, regardless of the speaker (analysis of visual materials will be the next step). Such transcription software will help enamour qualitative researchers of technology; a greater luxury for those who have transcribed interview tapes for analysis can hardly be imagined.

The real challenge, however, is in *analytic* software. My hope is that technology can help us think better . . . Although there are a number of computer programmes that claim to offer 'computer-assisted qualitative data analysis', such software does little more than neatly and systematically summarise material. While it allows for more material to be analysed, the quality of that analysis fully depends on the quality of the researcher/analyst's coding of the material. More material, more systematically summarised; this is what is currently on offer when it comes to technological assistance in data analysis. While the visuals that can be produced with some programmes look promising (e.g., offering a more accessible visualised overview), they could just as easily alienate non-academic participants in a project. With some software it is not even the danger of alienation or of rehearsing your own insight, but rather to move from the understanding and eventually theorisation of complex everyday talk, actions, gestures and visuals into quasi-quantitative summaries that appear to allow for the type of generalisation to populations that select sampled one-of-a-kind data can deliver but that theoretical sampling cannot.

What we are looking for is technology that can build high-quality matrices, such as the array of possibilities discussed by Miles and Huberman (1982), or produce spider visuals of how different (groups of) actors relate to single problematics from different perspectives, a positioning tool used in Dutch public broadcasting with quantitative materials. I would like to have technology—the computer in this case—work for us, much in the way that it allows for learning of complex relations in SPSS and such. Software that can digest complex interfaces and articulations, that can work with different levels of input: interview transcripts, observations and researcher field notes and memos, emerging theory and tentative models. All these aspects are part of qualitative data analysis and are at the root of its richness and openness to theorisation and the understanding of social and cultural practices. Just imagine your qualitative research assistant using the type of logarithms Amazon.com uses to tell you that interviewees who used these particular terms also liked to draw on a particular type of example or declined to answer negatively posed questions. 'Dear Joke Hermes, based on your track record and publications, your Qualitative Research Assistant 4.0 would like to suggest that you go back to material you collected in 2010 on this database, as you will find corroboration of your current tentative conclusion that cultural protectionism accords surprisingly well with neoliberal economics. The broad occurrence of this closed mindset, however, is not beneficial to your analysis reaching a wide audience. You may wish to reconsider the politics of your line of thought.'

TOWARDS A CONCLUSION

Threading together the different arguments, utopian wishes and fears, I find that I prefer not to think of technology and culture as an easy or even very logical combination—even though, in media culture (as perhaps elsewhere) they often are. To make headway in developing sensitive analytic software for qualitative research, new types of collaboration are needed between researchers and software developers. We are in need of a little technical training and a lot of utopian optimism—mixed, I should say, with some distrust and critical engagement with research practice. Most of all, this is needed in our capacities as editors and peer reviewers of each other's work, especially when it concerns the publication of research materials. We can and should be much more focused on the thoughtful, precise and intelligent discussion of methods used and decisions taken, and demand that authors reconstruct their process of making sense of how others make sense of media culture, texts and technology. Good research, and a responsible documentation of that research, can and should function as a bridge between everyday life, the academy and open practices of cultural criticism. As self-elected representatives of the people, we might at the very least put effort into justifying our choices and conclusions. As I am writing this, I realise this is very 'research 1.0'. But then, I expect it to take a while before we can move to a 2.0 version that is fully platform based. Luckily, even for research 1.0, the maxim that it gets better the more it is used is already true.

Technology as such will not set us free, but it may well help in opening up the different stages of research and data interpretation to those we work with. It may also be of great help in making qualitative data analysis a less opaque and more open practice. Would it not be a godsend if we had more access to the vast amount of data collected worldwide, provided we had good translation software, of course? As an audience researcher, I find I am not only interested in furthering media literacy and critical reflection, but would also like research practice to help with this. As this is not just the world of 2.0 media but also, in many ways, Guy Debord's (1967) 'society of the spectacle', technology will need to be tamed first and then put to use to help make interactive research and participative design 'fun'.

REFERENCES

Ang, I. 1990. Culture and Communication. Towards an Ethnographic Critique of Media Consumption in the Transnational Media System. *European Journal of Communication*, 5 (2–3), 239–260.

Berriman, L. 2009. *Re-Aligning Media and Actors: The Impact of 'Cross-Platform Broadcasting' in the Children's Media Industry*. Paper given at 'Ends of Television' conference, Amsterdam, June 2009.

Bovens, M. and P. 't Hart. 1996. *Understanding Policy Fiascoes*. New Brunswick, NJ: Transaction Publishers.

Butler, J. 1990. *Gender Trouble*. New York: Routledge.

Caldwell, J. T. 1995. *Televisuality: Style, Crisis, and Authority in American Television*. New Brunswick, NJ: Rutgers University Press.

Debord, G. 1967. *The Society of the Spectacle*. Available at http://www.marxists.org/reference/archive/debord/society.htm. Last accessed 1 October 2010.

Drotner, K. 1993. Media Ethnography: An Other Story. In *Nordisk forskning om kvinnor och medier*, edited by Ulla Carlsson. Göteborg: Nordicom, 25–40.

Gray, A. 1992. *Video Playtime: The Gendering of a Leisure Technology*. London: Routledge.

Gray, A. 2003. *Research Practice for Cultural Studies*. London; Sage.

Hammersley, M. and P. Atkinson. 1983. *Ethnography. Principles in Practice*. London: Tavistock.

Hartley, J. 1999. *Uses of Television*. London: Routledge.

Hermes, J. 2006. Hidden Debates: Rethinking the Relationship between Popular Culture and the Public Sphere. *Javnost/the Public*, 13(4), 27–44.

Hermes, J. 2009. Audience Studies 2.0. On the Theory, Politics and Method of Qualitative Audience Research. *Interactions: Studies in Communication and Culture*, 1(1), 111–127.

Jenkins, H. 2003. *Quentin Tarantino's Star Wars?: Digital Cinema, Media Convergence, and Participatory Culture*. Available at http://web.mit.edu/cms/People/henry3/starwars.html. Last accessed 2012-03-06.

Jenkins, H. 2006. *Convergence Culture. Where Old and New Media Collide*. New York: New York University Press.

Jensen, J. 1990. *Redeeming Modernity: Contradictions in Media Criticism*. London: Sage.

Johnson, S. 2005. *Everything Bad Is Good for You: How Today's Popular Culture Is Actually Making Us Smarter*. New York: Riverhead Books.

Johnson, S. 2010. *Where Good Ideas Come From: The Natural History of Innovation*. New York: Allen Lane.

Keen, A. 2007. *The Cult of the Amateur: How Today's Internet Is Killing Our Culture*. New York: Doubleday.

Kittler, F. 1997. *Gramophone, Film, Typewriter*. Stanford, CA: Stanford University Press.

Mander, J. 1978. *Four Arguments for the Elimination of Television*. New York: William Morrow.

Merton, R. K. 1949/1968. *Social Theory and Social Structure*. New York: Free Press.

McLuhan, M. 1964. *Understanding Media: The Extensions of Man*. New York: McGraw-Hill.

McRobbie, A. 1982. The Politics of Feminist Research: Between Talk, Text and Action. *Feminist Review*, 12, 46–57.

Miles, M. B. and A. M. Huberman. 1982. *Qualitative Data Analysis*. London: Sage

Morley, D. 1986. *Family Television*. London: Comedia

Peters, C. 2011. *TV Flat. Over de mediageletterdheid van kinderen*. MA thesis, Department of Media Studies, University of Amsterdam.

Postman, N. 1985. *Amusing Ourselves to Death: Public Discourse in the Age of Show Business*. New York: Penguin.

Shirky, C. 2008. *Here Comes Everybody: The Power of Organizing Without Organizations*. New York: Penguin Press.

Surowiecki, J. 2004. *The Wisdom of Crowds: Why the Many Are Smarter Than the Few and How Collective Wisdom Shapes Business, Economies, Societies and Nations*. New York: Doubleday.

Van Dijck, J. 2007. *Mediated Memories in the Digital Age*. Stanford, CA: Stanford University Press.

Wieten, J. 1994. Television's False Dawn. *Gazette* 53(3), 207–228.

Zielinski, S. 1999. *Audiovisions: Cinema and Television as Entr'actes in History*. Amsterdam: Amsterdam University Press.

Contributors

Maria Bakardjieva is professor in communication at University of Calgary, Canada. She is the author of *Internet Society: The Internet in Everyday Life* (Sage, 2005) and co-editor of *How Canadians Communicate* (U of Calgary P, 2004/2007). Her research has examined Internet use in everyday life across different social and cultural contexts. She has also published on the topics of online community, research ethics and e-learning. Her current projects focus on new media and citizens' participation in the public sphere. She is the editor of *Journal of Computer-Mediated Communication.*

Stina Bengtsson is senior lecturer in media and communication studies at Södertörn University, Sweden. She has a background in audience and everyday life studies, recently focussing on the cultural and societal dimensions of digital platforms such as virtual worlds. Her international publications include *The Don Quixote of Youth Culture* (Södertörn Academic Studies, 2005) and articles in journals such as the *Nordicom Review, International Journal of Cultural Studies* and *Journal of Virtual Worlds Research.*

Mats Björkin is senior lecturer in film studies at Göteborg University, Sweden. His research is focused on media history (industrial films, public relations, commercial films, television, distribution) and analysis of contemporary changes of film and television production. Recent publications include the 'Industrial Films and Cybernetics', in *Films That Work: Industrial Film and the Productivity of Media* (Amsterdam UP, 2009), and the chapter 'European Television Audiences: Localising the Viewers', in *A European Television History* (Blackwell, 2008). He is currently finishing a book on the industrial media landscape in Sweden during the 1950s.

Göran Bolin is professor of media and communication studies at Södertörn University, Sweden. He is the director of the research initiative Cultural Technologies. His recent international publications include *Value and the Media: Cultural Production and Consumption in Digital Markets*

(Ashgate, 2011), as well as articles in journals such as *Media, Culture & Society*, *New Media & Society*, *Popular Communication* and *Journal of Cultural Economy*.

Sandra Braman is professor of communication at the University of Wisconsin-Milwaukee, US. She has been studying the macro-level effects of the use of new information technologies and their policy implications since the mid-1980s. Her books include *Change of State: Information, Policy, and Power* (MIT Press, 2007/2009) and the edited volumes *Communication Researchers and Policy-Makers* (MIT Press, 2003), *The Emergent Global Information Policy Regime* (Palgrave Macmillan, 2004) and *Biotechnology and Communication: The Meta-Technologies of Information* (Lawrence Erlbaum Associates, 2004).

Elizabeth Van Couvering is a lecturer in media and communications at the University of Leicester, UK. Her interests in the political economy of new media stem, in part, from her experiences as a pioneer in the UK Internet industry. Publications include articles in the *Journal of Computer-Mediated Communication* and chapters in books such as *The Political Economies of the Media* (Bloomsbury Academic, 2011), *Web Search* (Springer, 2008) and *The Power of Search Engines* (Herbert von Halem, 2007).

Georgia Gaden is a PhD candidate at the Faculty of Communication at the University of Calgary, Canada. She has a background in feminist approaches to new media, and her doctoral research project is concerned with the relationship between understandings of textual value and quality of life for bloggers. Her work is published in journals such as *Journal of Virtual Worlds Research* and *Philosophy and Technology*.

Joke Hermes is reader in media, culture and citizenship at Inholland University, The Netherlands. She has published widely in the area of popular culture, with an emphasis on gender and audience studies. Among other works she is the author of *Reading Women's Magazines* (Polity, 1995) and *Re-Reading Popular Culture* (Wiley-Blackwell, 2005). She has also published widely in journals such as *Javnost/The Public*, *European Journal of Communication*, *Television and New Media* and *International Journal of Cultural Studies*. She is founder and co-editor of the *European Journal of Cultural Studies*.

Peter Jakobsson is lecturer in media and communication studies at Södertörn University, Sweden. His international publications include the co-authored chapter 'Googleplex and Informational Culture', in *Media Houses: Architecture, Media and the Production of Centrality* (Peter Lang, 2010), and articles in *European Journal of Cultural Studies*, *First Monday* and *Platform: Journal of Media & Communication*.

Toby Miller is distinguished professor of media and cultural studies at the University of California, Riverside, US, and directs the University of California's study-abroad program in Mexico. His latest books are *Cultural Citizenship: Cosmopolitanism, Consumerism and Television in a Neoliberal Age* (Temple UP, 2007), *Makeover Nation: The United States of Reinvention* (Ohio State UP, 2008), the *Contemporary Hollywood Reader* (Routledge, 2009) and *Television Studies: The Basics* (Routledge, 2010). *Greening the Media* (with Richard Maxwell) is in press with Oxford. He is the editor of the journal *Social Identities* (with Pal Ahluwalia) and runs the podcast 'culturalstudies', available free on iTunes and through applications for Android and iPhone.

Lisa Parks is professor of film and media studies at University of California, Santa Barbara, US. She is the author of *Cultures in Orbit: Satellites and the Televisual* (Duke UP 2005), and co-editor of *Planet TV* (NYU Press 2003), *Undead TV* (Duke UP, 2007) and *Down to Earth: Satellite Technologies, Industries and Cultures* (Rutgers UP, 2012). She is currently working on two new books: *Coverage: Media Space and Security after 9/11* (Routledge, forthcoming) and *Mixed Signals: Media Infrastructures and Cultural Geographies.*

Eric W. Rothenbuhler is professor of media arts and studies, and associate dean for research and graduate education in the Scripps College of Communication, Ohio University, US. He is the author of *Ritual Communication* (Sage, 1998), co-editor of *Media Anthropology* (Sage, 2005) and *Communication and Community* (Lawrence Erlbaum, 2001), and author or co-author of over sixty articles, chapters, essays and reviews on media, ritual, community, media industries, popular music and communication theory.

Fredrik Stiernstedt is lecturer in media and communication studies at Södertörn University, Sweden. His recent international publications include the co-authored chapter 'Googleplex and Informational Culture', in *Media Houses: Architecture, Media and the Production of Centrality* (Peter Lang, 2010) and articles in *The Radio Journal*, *First Monday* and *Participations.*

William Uricchio is professor and director of comparative media studies at the Massachusetts Institute of Technology (MIT), US, and professor of comparative media history at Utrecht University, The Netherlands. His research focuses on the emergence of media technologies and practices, and recent books include *We Europeans? Media, Representations, Identity* (Intellect, 2008) and *Media Cultures* (Universitätsverlag Winter, 2006).

Index

A
Amazon 5, 21, 34, 199
Aristotle 1, 2
art 1–4, 9, 51, 124, 177, 190–2
art project 30
artist 27, 60, 74, 78, 126
artworks 33
digital arts 51
audience/s
active audience theory 12, 88, 96
audience commodity 7
audience research 13, 189, 192, 194, 195, 198
control of (*see also* surveillance) 87
young audiences 92, 93
audio 38
augmented reality 19, 20, 23, 29, 30 32, 33

B
Berland, Judy 9, 21
Bing 12, 118
Brand Estonia 6, 184

C
Canaletto 27–29, 32
capitalism 126, 172
corporate 88
digital 162
global 125
laissez-faire 7
late 7
multinational 19
capitalist 96, 126, 161, 162
Carey, James 64
communication 7, 11, 49
communication infrastructures 64
communication technologies 1, 7, 9
mediated 7
networks of 7
transnational flows of 11
audio communication 38,
consumption 6, 53, 58, 68, 92, 96, 162, 191
citizenship and 88
culture of 129
logic of 48
means of 8
media 51, 57, 87, 172,
convergence 38, 139, 161
convergence culture 57, 160, 189, 197
Crystal Palace 8
Culler, Jonathan 19–23, 32–3

D
Descartes 20, 21, 28, 29
digital
arts 51
commodities 46
culture 103, 109
economy 161–2
images 25
information 12, 103, 114
media 8, 48, 53, 54, 103, 170, 171
recording 46–7,
technology 4, 11, 108, 140
digitization 8, 11

E
Estonia 6, 13, 170
Estonian Art Museum, see also KuMu 176
Estonian government 6, 13
Estonian embassy 13, 170–1, 173–86
European Community, EC 5
Eurovision Song Contest 6, 173
Eesti stiil 6

F

Facebook 7, 12, 103, 113, 118, 161, 164, 166, 169
file-sharing 11, 51–2, 54–5, 57, 59, 61
Fiske, John 64
Frankfurt School 140

G

Gogh, Teo Van 75
Google 12, 21, 103, 109, 114, 118, 127
Google Street View 26, 32,
Google AdSense 34

H

Hall, Stuart 96
Heidegger, Martin 1, 2, 10, 20, 21, 28, 29, 31, 34, 136, 140
Hollywood 53, 55, 87,

I

industrial exhibitions 2
information
  commercial 88
  dissemination of 7, 11, 12
  digital 12, 103, 114,
  flow of 142
  Freedom of Information Act 77
  photographic 28
  retrieval 121, 125
  policy 133, 134, 136–8, 143, 145
  storage 108
  virtual 33
  visual 27,
  web-based 30
information society 103, 139
Innis, Harold 64
Internet
  access 139
  companies 106,
  culture 127–8, 130
  traffic 8
Internet Movie Database 57
Invest Sweden 107
Interbrand 6

J

James Bond 3, 111

K

Kittler, Friedrich 8–9
KuMu (Estonian Art Museum) 176

L

Langer, František 51–4
Locke, John 12, 133–4, 137–44
London 8
Luddism 4

M

Marx, Karl 113
means of consumption 8
means of production 7, 134
McLuhan, Marshall 8
Mechanical Turk 5
Microsoft 20, 24, 89, 103, 118
Monnet, Jean 5
Mumford, Lewis 2
music 3, 7, 11, 36–49, 54, 56, 88, 160, 176
musicians 25

N

nation branding 5–6, 13, 173, 180, 184
nation-state 6, 170–2, 184–5
Nicomachean Ethics (Aristotle) 1
No Sense of Place (Meyrowitz) 4
North of Empire (Berland) 9

P

participation 43, 92, 130, 144
peer-to-peer 52, 54–6, 58–62
Pirate Bay, The 54, 58, 59
Potrč, Marjetica 74
PR 6
production
  audio 45
  culture of 129
  film 51
  processes 5, 133
  media 8, 191,
  music 11
  of facts 138
  of space 9, 171, 184
  of knowledge 10, 141, 144,
  technologies of 154, 160, 161, 165,
  search, search engine 118, 120, 124, 128
production force 6

R

Ruin, Hans 1, 2

S

Shakespeare, William 114
Schivelbush, Wolfgang 6
Seadragon technology 24
search engines 12
Second Life 13

Semiotics of Tourism (Culler) 19
Silent Running 112
smart phone 30
Snavley, Noah 24
surveillance 8, 11, 12, 75, 76, 87–8, 90, 92, 94–5, 98–9, 141
Svanberg, Lasse 3
Swedish Christian Democratic Party 107

T

technē 1–2, 190
techno-sceptic, technofobic 3, 4
technology
  and knowledge 11, 136
  audio 39, 42
  computer 111
  concept of 1, 157, 190
  consumer 36
  culture and 9
  cultures of 2–5, 13
  cultural 5, 8, 13, 136, 145n, 189
  digital 11, 108
  image 29
  material 1
  media 7, 8 , 173, 191, 193, 195–8
  policy 133, 141
  portal 69, 70, 80
television 7
Television: Technology and Cultural Form (Williams) 8
time
  historical 107–10
  technological 110–113
twitter 12

U

Understanding Media: The Extensions of Man (McLuhan) 8

W

Web 2.0 8, 13, 154, 157, 159–62, 164–6, 173, 197–8
Wikipedia 19, 27, 34, 89, 143
Wilders, Geert 75
Williams, Raymond 8
World's Fairs 2–3

Y

Yahoo! 12

An environmentally friendly book printed and bound in England by www.printondemand-worldwide.com

PEFC Certified

This product is from sustainably managed forests and controlled sources

www.pefc.org

PEFC/16-33-415

MIX
Paper from responsible sources
FSC® C004959
www.fsc.org

This book is made entirely of chain-of-custody materials; FSC materials for the cover and PEFC materials for the text pages.

#0046 - 170113 - C0 - 229/152/12 [14] - CB